THE CREATION

ALSO BY EDWARD O. WILSON

The Theory of Island Biogeography,
with Robert H. MacArthur (1967)

A Primer of Population Biology, with William H. Bossert (1971)

The Insect Societies (1971)

Sociobiology: The New Synthesis (1975)

On Human Nature (1978)

Caste and Ecology in the Social Insects,
with George F. Oster (1978)

Genes, Mind, and Culture, with Charles J. Lumsden (1981)

Promethean Fire, with Charles J. Lumsden (1983)

Biophilia (1984)

The Ants, with Bert Hölldobler (1990)

*Success and Dominance in Ecosystems:
The Case of the Social Insects* (1990)

The Diversity of Life (1992)

Journey to the Ants, with Bert Hölldobler (1994)

Naturalist (1994)

In Search of Nature (1996)

Consilience: The Unity of Knowledge (1998)

Biological Diversity: The Oldest Human Heritage (1999)

The Future of Life (2002)

Pheidole in the New World (2003)

*From So Simple a Beginning:
The Four Great Books of Charles Darwin* (2005)

Nature Revealed: Selected Writings, 1949–2006 (2006)

THE CREATION

AN APPEAL TO SAVE LIFE ON EARTH

Edward O. Wilson

W·W·Norton & Company NEW YORK LONDON

Manufacturing by RR Donnelley, Harrisonburg
Book design by Margaret Wagner
Production manager: Julia Druskin

ISBN 13: 978-0-393-06217-5
ISBN 10: 0-393-06217-1

W. W. Norton & Company, Inc.
500 Fifth Avenue, New York, N.Y. 10110
www.wwnorton.com

W. W. Norton & Company Ltd., Castle House,
75/76 Wells Street, London W1T 3QT

1 2 3 4 5 6 7 8 9 0

Contents

THE CREATION

DECLINE AND REDEMPTION

III

WHAT SCIENCE HAS LEARNED

IV

TEACHING THE CREATION

V

REACHING ACROSS

THE CREATION

I

The Creation

A CALL FOR HELP

AND AN INVITATION TO VISIT

THE EMBATTLED NATURAL WORLD

IN THE COMPANY OF

A BIOLOGIST

1

Letter to a Southern Baptist Pastor: Salutation

DEAR PASTOR:

We have not met, yet I feel I know you well enough to call you friend. First of all, we grew up in the same faith. As a boy I too answered the altar call; I went under the water. Although I no longer belong to that faith, I am confident that if we met and spoke privately of our deepest beliefs, it would be in a spirit of mutual respect and good will. I know we share many precepts of moral behavior. Perhaps it also matters that we are both Americans and, insofar as it might still affect civility and good manners, we are both Southerners.

I write to you now for your counsel and help. Of course, in doing so, I see no way to avoid the fundamental differences in our respective worldviews. You are a literalist interpreter of Christian Holy Scripture. You reject the conclusion of science that mankind evolved from lower forms. You believe that each person's soul is immortal, making this planet a way station to a second, eternal life. Salvation is assured those who are redeemed in Christ.

I am a secular humanist. I think existence is what we make of it as individuals. There is no guarantee of life after death, and

3

heaven and hell are what we create for ourselves, on this planet. There is no other home. Humanity originated here by evolution from lower forms over millions of years. And yes, I will speak plain, our ancestors were apelike animals. The human species has adapted physically and mentally to life on Earth and no place else. Ethics is the code of behavior we share on the basis of reason, law, honor, and an inborn sense of decency, even as some ascribe it to God's will.

For you, the glory of an unseen divinity; for me, the glory of the universe revealed at last. For you, the belief in God made flesh to save mankind; for me, the belief in Promethean fire seized to set men free. You have found your final truth; I am still searching. I may be wrong, you may be wrong. We may both be partly right.

Does this difference in worldview separate us in all things? It does not. You and I and every other human being strive for the same imperatives of security, freedom of choice, personal dignity, and a cause to believe in that is larger than ourselves.

Let us see, then, if we can, and you are willing, to meet on the near side of metaphysics in order to deal with the real world we share. I put it this way because you have the power to help solve a great problem about which I care deeply. I hope you have the same concern. I suggest that we set aside our differences in order to save the Creation. The defense of living Nature is a universal value. It doesn't rise from, nor does it promote, any religious or ideological dogma. Rather, it serves without discrimination the interests of all humanity.

Pastor, we need your help. The Creation—living Nature—is in deep trouble. Scientists estimate that if habitat conversion and other destructive human activities continue at their present rates, half the species of plants and animals on Earth could be either gone or at least fated for early extinction by the end of the century. A full quarter will drop to this level during the next half century as a result of climate change alone. The ongoing extinction rate is calculated in the most conservative estimates to be about a hundred times above that prevailing before humans appeared on Earth, and it is expected to rise to at least a thousand times greater or more in the next few decades. If this rise continues unabated, the cost to humanity, in wealth, environmental security, and quality of life, will be catastrophic.

Surely we can agree that each species, however inconspicuous and humble it may seem to us at this moment, is a masterpiece of biology, and well worth saving. Each species possesses a unique combination of genetic traits that fits it more or less precisely to a particular part of the environment. Prudence alone dictates that we act quickly to prevent the extinction of species and, with it, the pauperization of Earth's ecosystems—hence of the Creation.

You may well ask at this point, Why me? Because religion and science are the two most powerful forces in the world today, including especially the United States. If religion and science could be united on the common ground of biological conservation, the problem would soon be solved. If there is any moral precept shared by people of all beliefs, it is that we owe ourselves and future generations a beautiful, rich, and healthful environment.

I am puzzled that so many religious leaders, who spiritually represent a large majority of people around the world, have hesitated to make protection of the Creation an important part of their magisterium. Do they believe that human-centered ethics and preparation for the afterlife are the only things that matter? Even more perplexing is the widespread conviction among Christians that the Second Coming is imminent, and that therefore the condition of the planet is of little consequence. Sixty percent of Americans, according to a 2004 poll, believe that the prophecies of the book of Revelation are accurate. Many of these, numbering in the millions, think the End of Time will occur within the life span of those now living. Jesus will return to Earth, and those redeemed by Christian faith will be transported bodily to heaven, while those left behind will struggle through severe hard times and, when they die, suffer eternal damnation. The condemned will remain in hell, like those already consigned in the generations before them, for a trillion trillion years, enough for the universe to expand to its own, entropic death, time enough for countless universes like it afterward to be born, expand, and likewise die away. And that is just the beginning of how long condemned souls will suffer in hell— all for a mistake they made in choice of religion during the infinitesimally small time they inhabited Earth.

For those who believe this form of Christianity, the fate of ten million other life forms indeed does not matter. This and other similar doctrines are not gospels of hope and compassion. They are gospels of cruelty and despair. They were not born of the heart of Christianity. Pastor, tell me I am wrong!

However you will respond, let me here venture an alternative ethic. The great challenge of the twenty-first century is to raise people everywhere to a decent standard of living while preserving as much of the rest of life as possible. Science has provided this part of the argument for the ethic: the more we learn about the biosphere, the more complex and beautiful it turns out to be. Knowledge of it is a magic well: the more you draw from it, the more there is to draw. Earth, and especially the razor-thin film of life enveloping it, is our home, our wellspring, our physical and much of our spiritual sustenance.

I know that science and environmentalism are linked in the minds of many with evolution, Darwin, and secularism. Let me postpone disentangling all this (I will come back to it later) and stress again: to protect the beauty of Earth and of its prodigious variety of life forms should be a common goal, regardless of differences in our metaphysical beliefs.

To make the point in good gospel manner, let me tell the story of a young man, newly trained for the ministry, and so fixed in his Christian faith that he referred all questions of morality to readings from the Bible. When he visited the cathedral-like Atlantic rainforest of Brazil, he saw the manifest hand of God and in his notebook wrote, "It is not possible to give an adequate idea of the higher feelings of wonder, admiration, and devotion which fill and elevate the mind."

That was Charles Darwin in 1832, early into the voyage of HMS *Beagle*, before he had given any thought to evolution.

And here is Darwin, concluding *On the Origin of Species* in 1859, having first abandoned Christian dogma and then, with his newfound intellectual freedom, formulated the theory of evolution by natural selection: "There is grandeur in this view of life, with its several powers, having been originally breathed into a few forms or into one; and that, whilst this planet has gone cycling on according to the fixed law of gravity, from so simple a beginning endless forms most beautiful and most wonderful have been, and are being, evolved."

Darwin's reverence for life remained the same as he crossed

the seismic divide that divided his spiritual life. And so it can be for the divide that today separates scientific humanism from mainstream religion. And separates you and me.

You are well prepared to present the theological and moral arguments for saving the Creation. I am heartened by the movement growing within Christian denominations to support global conservation. The stream of thought has arisen from many sources, from evangelical to unitarian. Today it is but a rivulet. Tomorrow it will be a flood.

I already know much of the religious argument on behalf of the Creation, and would like to learn more. I will now lay before you and others who may wish to hear it the scientific argument. You will not agree with all that I say about the origins of life— science and religion do not easily mix in such matters—but I like to think that in this one life-and-death issue we have a common purpose.

2

Ascending to Nature

AT THE VERY LEAST, Pastor, I expect we agree that somehow and somewhere back in history humanity lost its way. As a Christian minister, you will likely respond that of course we lost our way, we departed from Eden. Our progenitors made a terrible mistake, and so we live in original sin. Now we wander between heaven and hell, above the animals but below the angels, as we await ascension to a better world through faith in the Redeemer.

Would you be willing to suppose that part of Eden was the rest of life as it was before humanity? The book of Genesis affirms that much, whether read literally or metaphorically. The conclusion of science also is that such a primordial world existed and served as the cradle of humanity. Yet—if biology has learned anything, it is that our species did not, in contradiction to a literalist reading of Genesis, come abruptly into existence by a touch of divine fire. Instead, we evolved in a biologically rich world over tens of thousands of generations. Nor were we driven from this Eden. Instead, we destroyed most of it in order to improve our lives and generate more people. Billions of more people, to the peril of the Creation. I would like to offer the following explanation of the human dilemma:

According to archaeological evidence, we strayed from Nature with the beginning of civilization roughly ten thousand years ago. That quantum leap beguiled us with an illusion of freedom from the world that had given us birth. It nourished the belief that the human spirit can be molded into something new to fit changes in the environment and culture, and as a result the timetables of history desynchronized. A wiser intelligence might now truthfully say of us at this point: here is a chimera, a new and very odd species come shambling into our universe, a mix of Stone Age emotion, medieval self-image, and godlike technology. The combination makes the species unresponsive to the forces that count most for its own long-term survival.

There seems no better way to explain why so many smart people remain passive while the precious remnants of the natural world disappear. They are evidently unaware that ecological services provided scott-free by wild environments, by Eden, are approximately equal in dollar value to the gross world product. They choose to remain innocent of the historical principle that civilizations collapse when their environments are ruined. Most troubling of all, our leaders, including those of the great religions, have done little to protect the living world in the midst of its sharp decline. They have ignored the command of the Abrahamic God on the fourth day of the world's birth to "let the waters teem with countless living creatures, and let birds fly over the land across the vault of heaven."

I hesitate to introduce a beautiful subject with an animadversion. Few will deny, however, that the human impact on the natural environment is accelerating and makes a frightening picture.

What are we to do? At the very least, put together an honest history, one on which people of many faiths can in principle agree. If such can be fashioned, it will serve at least as prologue to a safer future.

We can begin with the key discovery of green history: *Civilization was purchased by the betrayal of Nature*. The Neolithic revolution, comprising the invention of agriculture and villages, fed on Nature's bounty. The forward leap was a blessing for humanity. Yes, it was: those who have lived among hunter-gatherers will tell you they are not at all to be envied. But the revolution encouraged the false assumption that a tiny selection of domesticated plants and animals can support human expansion indefinitely. The pauperization of Earth's fauna and flora was an acceptable price until recent centuries, when Nature seemed all but infinite, and an enemy to explorers and pioneers. The wildernesses and the aboriginals surviving in them were there to be pushed back and eventually replaced, in the name of progress and in the name of the gods too, lest we forget.

History now teaches a different lesson, but only to those who will listen. Even if the rest of life is counted of no value beyond the satisfaction of human bodily needs, the obliteration of Nature is a dangerous strategy. For one thing, we have become a species specialized to eat the seeds of four kinds of grass—wheat, rice, corn, and millet. If these fail, from disease or climate change, we too shall fail. Some fifty thousand wild plant species (many of which face extinction) offer alternative food sources. If one insists on being thoroughly practical about the matter, allowing these and the rest of wild species to exist should be considered part of a portfolio of long-term investment. Even the most recalcitrant people must come to view conservation as simple prudence in the management of Earth's natural economy. Yet few have begun to think that way at all.

Meanwhile, the modern technoscientific revolution, including especially the great leap forward of computer-based information technology, has betrayed Nature a second time, by fostering the belief that the cocoons of urban and suburban material life are sufficient for human fulfillment. That is an especially serious mistake. Human nature is deeper and broader than the artifactual contrivance of any existing culture. The spiritual roots of *Homo sapiens* extend deep into the natural world through still mostly hidden channels of mental development. We will not reach our full potential without understanding the origin and hence meaning of the aesthetic and religious qualities that make us ineffably human.

Granted, many people seem content to live entirely within the synthetic ecosystems. But so are domestic animals content, even in the grotesquely abnormal habitats in which we rear them. This in my mind is a perversion. It is not the nature of human beings to be cattle in glorified feedlots. Every person deserves the option to travel easily in and out of the complex and primal world that gave us birth. We need freedom to roam across land owned by no one but protected by all, whose unchanging horizon is the same that bounded the world of our millennial ancestors. Only in what remains of Eden, teeming with life forms independent of us, is it possible to experience the kind of wonder that shaped the human psyche at its birth.

Scientific knowledge, humanized and well taught, is the key to achieving a lasting balance in our lives. The more biologists learn about the biosphere in its full richness, the more rewarding the image. Similarly, the more psychologists learn of the development of the human mind, the more they understand the gravitational pull of the natural world on our spirit, and on our souls.

We have a long way to go to make peace with this planet, and with each other. We took a wrong turn when we launched the Neolithic revolution. We have been trying ever since to ascend *from* Nature instead of *to* Nature. It is not too late for us to come around, without losing the quality of life already gained, in order to receive the deeply fulfilling beneficence of humanity's natural heritage. Surely the reach of religious belief is great enough, and its teachers generous and imaginative enough, to encompass this larger truth not adequately expressed in Holy Scripture.

Part of the dilemma is that while most people around the world care about the natural environment, they don't know why they care, or why they should feel responsible for it. By and large they have been unable to articulate what the stewardship of Nature means to them personally. This confusion is a great problem for contemporary society as well as for future generations. It is linked to another great difficulty, the inadequacy of science education, everywhere in the world. Both arise in part from the explosive growth and complexity of modern biology. Even the best scientists have trouble keeping up with more than a small part of what has emerged as the most important science for the twenty-first century.

I believe that the solution to all of the three difficulties—ignorance of the environment, inadequate science education, and the bewildering growth of biology—is to refigure them into a single problem. I hope you will agree that every educated person should know something about the core of this unified issue. Teacher and student alike will benefit from a recognition that living Nature has opened a broad pathway to the heart of science itself, that the breath of our life and our spirit depend upon its survival. And to grasp and discuss on common ground this principle: because we are part of it, the fate of the Creation is the fate of humanity.

3

What Is Nature?

DO YOU AGREE, PASTOR, that the depth and the complexity of living Nature still exceed human imagination? If God seems unknowable, so too does most of the biosphere. Biologists never cease to stress how little we understand of the living world around us. Domestic breeds of plants and animals are but trivial variants within the diversity of life. Our most sophisticated simulations of life processes still fall short of the real thing. We cannot yet create an artificial organism at even the lowest level. New worlds and endless discoveries await in Nature, among them the solution to that mystery of mysteries, the meaning of human life.

But what is Nature? The simplest possible answer is also the best: Nature is that part of the original environment and its life forms that remains after the human impact. Nature is all on planet Earth that has no need of us and can stand alone.

Some skeptics have insisted that even when elaborated, such a definition has little use, because the natural world has been so disturbed as to be humanized everywhere and thus has lost its original identity. There is a kernel of truth in that claim. Very few square kilometers of Earth's land surface remain untrodden at some time or other, at the very least by explorers and native people. In 1955 I was the first non-Papuan to reach the summit ridge of the central Sarawaget Mountains of north-eastern New Guinea. (Admittedly, few if any others had tried, and I was still young enough to think myself invulnerable.) After four days of toiling upward through virgin midmountain and cloud forest, discovering new species of ants and frogs along the way, I proudly deposited a bottle with the record of my achievement inside a rock-pile cairn on the crest. But I was led to that spot by native hunters who often visited the area in search of alpine wallabies, plump little kangaroos that abound in the tussock grassland above the tree line. I've often wondered how many times my companions, and their forebears for thousands of years on back, had been there already, and by what sylvan routes they had traversed the forest to reach this particular spot. Surely a lot of people, with a deep and rich history to them.

It is further true that thousands of industrial pollutants drift continuously onto the receding polar snows and into the most distant seas. Five percent of Earth's land surface is burned every year, mostly in order to create agricultural fields or refertilize old ones. These practices and others contribute to the overloading of the atmosphere with greenhouse gases, sufficient to destabilize the climates of the entire planet.

The humanization of Earth proceeds in many other ways. Most of the land-dwelling megafauna, comprising animals weighing ten kilograms or more, have been hunted to extinction on the land. Wildlife of the plains and forests of the world today bears little resemblance to the majestic parade of giant mammals and birds driven to extinction by expert Paleolithic hunters. A large minority of those surviving today are on the endangered list. Twelve thousand years ago the wildlife of the American plains was richer than that of Africa today.

Overall, humanity has altered this planet as profoundly as our considerable powers permit. Yet a great deal of Nature does remain. In purest state it exists in what are still legitimately called wildernesses. Very roughly, a full-scale, megafauna-sized wilderness is defined as a relatively large and mostly undisturbed aggregate of contiguous habitats. As specified by Conservation International in a recent study, it is an expanse of ten thousand square kilometers (one million hectares) or more, at least 70 percent of whose area still bears natural vegetation. Domains of this magnitude include the great tropical forests of the Amazon Basin, the Congolian Basin, and most of the island of New Guinea. They also include the taiga, the belt of mostly coniferous forests that stretches across North America and on through Siberia to Finnoscandia. Wildernesses of a very different kind are Earth's largest deserts, the polar regions, the high seas, and the abyssal floors of the oceans (in contrast, very few deltas and coastal waters remain unchanged).

Smaller wildernesses abound, denoted officially in the U.S. Wilderness Act of 1964 as parts of Earth "untrammeled by man and where man himself is a visitor who does not remain." In this historic piece of legislation, 9.1 million acres were set aside "for the use and enjoyment of the American people in such a manner as will leave them unimpaired for future use and enjoyment." Mandating protection of fragments as small as 5,000 acres, the act has saved such priceless tracts of land and water as Montana's Great Bear Wilderness and the Allagash Wilderness Waterway of Maine.

Untrammeled. How well that word catches the spirit of wilderness! But how exactly it applies in practice depends on the scale employed. A suburban woodlot is obviously no longer a wilderness for mammals, birds, and trees. But it might be a "microwilderness" for small organisms. Many kinds of insects, mites, and other arthropods, mostly under ten millimeters in size, range freely there, their local domains untroubled by human hands, feet, or tools. Luckily, microwildernesses are not a trivial part of wild Nature. Quite the opposite: each cubic meter of soil and humus within it is a world swarming with hundreds of thousands of such creatures, representing hundreds of species. With them are even greater numbers and diversity of microbes. In one gram of soil, less than a handful, live on the order of ten billion bacteria belonging to as many as six thousand species.

The entire lives of the microscopic and barely visible organisms play out in spaces that human beings, among the largest animals on Earth, are inclined to dismiss. For an oribatid mite, only a crawling dot to the naked eye, a rotting tree stump is the equivalent of Manhattan. For a bacterium, the equivalent is New York State. The woodlot may be seriously disturbed on a macroscale, as perceived by humans who can walk across it in a few minutes. It may be littered with trash. Its trees may be second-growth. But around the base of each tree is an ancient and relatively intact world of miniature inhabitants. The soil and litter between the trees is their continent, and the nearby springtime pool is their sea.

The idea of the microwilderness is one of the main reasons I recently became interested in the Boston Harbor Islands national park area. The harbor had been in continuous heavy use since the mid-1600s, and for almost all that time had also

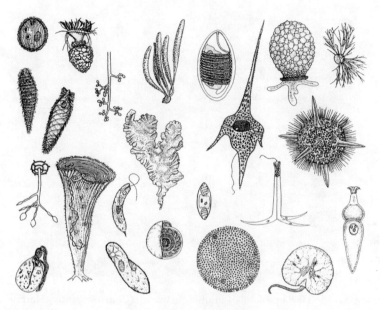

Microscopic inhabitants of microwildernesses. Depicted here are algae,
protozoans, and fungi. (From John O. Corliss, "Biodiversity and
Biocomplexity of the Protists and an Overview of Their Significant Roles
in Maintenance of Our Biosphere," *Acta Protozoologica* 41
[2002]: 199–219.)

served as a capacious municipal sewer. In 1985 its waters were
ranked the most polluted among all of America's harbors. Its
thirty-four grubby little islands were counted of little value to
New England's largest city, even though the nearest are only an
hour's rowboat trip away. During the 1990s, the situation
changed as wastewater outflow of Greater Boston was cleansed
by a new filtration system. The promise of the Harbor Islands as
a recreational area then became obvious, and their importance
in science and education grew.

Today, the archipelago, reborn as the Boston Harbor Islands

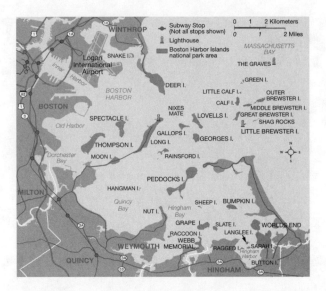

The Boston Harbor Islands National Park area. (Courtesy Boston Harbor Islands National Park.)

national park area, is a mecca for residents and visitors. The harbor waters offer proof of the resilience of living nature. Shellfish have resettled the bottom. Large fish are back: striped bass and bluefish run right up to the harbor docks. Seals and porpoises have returned in small numbers; even a humpback whale was observed cruising the outer island waters, presumably drawn there by a new abundance of food.

Because so much of my life's work had centered on the biology of islands, and often drew me to remote parts of the world, I was attracted to the prospect of a natural laboratory and classroom at my doorstep, one that also serves seven million other suburban and urban dwellers. Best of all, here was an opportunity to unglue city children from their televisions and computers and engage them in real-life educational adventure. It had the potential for hands-on introduction to science; and on the side, and not least, it could help offset the intimidating high-tech activity of Harvard and MIT close by. The message is this: first-rate science need not start with white coats and scribbles on a blackboard.

I admit to yet another, personal reason for my interest. My great-grandfather William C. Wilson, a Confederate blockade runner known to his friends as Black Bill, had been imprisoned in Fort Warren on George's Island after his capture in 1863 during a run at the entrance of Mobile Bay. On a pleasant morning in the fall of 2004, I visited his old cellblock and learned from an 1865 menu that, in company of his fellow prisoner Alexander Stevens, vice president of the Confederacy, he had lived surprisingly well for at least a short period following the war. He had arrived at the fort in poor health, having endured brutal conditions in two previous federal prisons. Black Bill's problem was that by federal law he was a common criminal—not an enemy naval officer, but a civilian bar pilot who had used his skills to run supplies from Cuba into the port of Mobile. Fort Warren was a maximum-security prison used to house naval officers and blockade runners, two categories considered by Secretary of War Stanton to pose an exceptional threat to the Union war effort. Black Bill drew an extra year at Fort Warren for insubordination (spitting on a guard, according to family oral record). He died in 1872, from an undiagnosed disease contracted during his earlier period of confinement.

How strange, then, it seemed to me upon arriving at the fort that Black Bill and I would cross paths in this improbable place, in two such improbable roles, a felon by happenstance of war followed by the entomologist to whom he had passed on one-eighth of his genetic code, there to study insects.

The Boston Harbor Islands draw naturalists to them in part because they support an intensely cosmopolitan flora and fauna. During constant heavy exposure for over three centuries to incoming seaborne traffic, they have been colonized by a large number of non-native plant and insect and other invertebrate animal species, mostly European in origin. Overall, for example, 229, or 44 percent, of the 521 plant species recently censused are exotic. These cargo stowaways and ship jumpers, some originating from populations that were first established on the surrounding mainland, today mingle with the native species to form complex assemblages. Larger animals—wildlife in the conventional sense—are also present. They consist mostly of seabirds and migratory land birds, in sufficient variety to attract serious birders from around New England and beyond.

The modest little archipelago takes on whole new meaning when microbes, fungi, and small invertebrates are added. Then the islands are seen as a world of unexplored microwildernesses. When portable microscopes are added—now widely available and relatively inexpensive—the discovery of microscopic and near-microscopic organisms can begin. Biodiversity surveys will be at last truly comprehensive. When the scientific exploration is made entertaining and combined with education, a new form of civic institution takes root.

Some postmodernist philosophers, convinced that truth is relative and dependent only on a person's worldview, argue that there is no such objective entity as Nature. It is, they say, a false dichotomy that has arisen in some cultures and not in other cultures. I am willing to entertain such a belief, for a few minutes anyway, but I have crossed too many sharp boundaries between natural and humanized ecosystems to doubt the objectivity of Nature.

I need not limit my account to the environs of Boston. You can, for example, share one of the most dramatic experiences I have had, many times over the years, by making a casual visit to the Florida Keys. You start with a journey down the commercial gauntlet of U.S. 1 to the Lower Keys. This strip is not the reality of southernmost Florida, not the residence of its ancient history and timeless spirit. In order to find that, stop at a boat rental office on the edge of the Great White Heron National Wildlife Reserve. Take a fourteen-footer out in the direction of the Gulf of Mexico and enter the channels that meander around the fringing red mangrove islets. Secure your boat at the edge of an islet with a higher mudflat base. Climb past the stilt roots of the outer trees. You are now in a fragment of a virgin forest. It has never been cut, because the wood has little or no commercial value and the tidal flats on which it stands cannot be otherwise developed. Its tangled growth is a nursery for the organisms of land and sea. The green vegetation and decaying tree branches teem with thousands of species of insects and other miniature wildlife. The shallow water lapping the peripheral roots supports an astonishing multitude of fish, shrimp and other crustaceans, anemones, and a legion of less familiar marine life forms. Much of the mangrove fauna is still unknown to science. Humanity's artifactual ecosystem, the commercial strip bounding the forest to the east, and the only one entered by the vast majority of visitors, is less than eighty years old. The mangrove forest, in more or less the same form it has today, the habitat skirted by unseeing visitors, has occupied parts of the

Gulf coast for millions of years. If humans were to abandon the Florida Keys, the humanized land would revert in a few decades to mudflats and mangrove islands probably indistinguishable from those that still survive.

If it is hard data you need to distinguish Nature from non-Nature, consider tropical rainforests. Although they cover only about 6 percent of the land surface of the planet—about the same as the forty-eight contiguous United States—they are the headquarters of Earth's terrestrial biodiversity, harboring over half of the known plant and animal species. There is a rule that all naturalists know and talk about who work in rainforests. The plant or animal species that catches your eye at this moment you may not see again that day, or week, or even year. It may never reveal itself to you again, no matter how long and hard you search. The tropical rainforest is home to a large number of such very rare and elusive life forms. Why this is the case is an enduring mystery that only now has begun to attract serious scientific study.

A stunning contrast exists between the rainforest and surrounding non-rainforest habitats cleared and developed by humans. In a few square kilometers at Jari, in the western Brazilian state of Rondônia, entomologists have recorded sixteen hundred kinds of butterflies. In nearby pastures of similar extent converted from rainforest by logging and burning, there may be (I know of no exact count but have looked around in similar places) fifty species, plus an indeterminate number that stray across the inhospitable terrain from one forest fragment to another. The same disproportion is true for mammals, birds, frogs, spiders, ants, beetles, fungi, and other organisms, including, with a vengeance, thousands of species of trees and countless life forms that dwell in the canopy.

I admit that in many other places the transition between Nature and not-Nature is not nearly so precipitous. The real, people-permeated world has been transformed into a kaleidoscope of extremes and intermediates, grading from still primal, million-year-old habitats all the way down to parking lots. The shift of this planetary kaleidoscope is toward the humanized, the simplified, the unstable.

But wait! Remember the microwildernesses. Nature dies hard. Even in the parking lot extremum, notice the resilient little weed that peeps from a crack in the concrete, the tuft of grass holding on at the curb, the faint colorous span of the cyanobacterial colony plastered next to the ticket kiosk. Look closely for tiny creatures that thrive in their parsimonious midst: the mite, the nematode worm, the caterpillar struggling to grow into a moth. These last-stand wild organisms, the vanguard of Earth's inevitable return to green and blue, wait patiently for us to change our mind. Their species are still able to give back some of what we remain so remorselessly bent on destroying.

4

Why Care?

I WILL ARGUE, PASTOR, that Nature is not only an objective entity but vital to our physical and spiritual well-being. I expect you'll concur with that, although your logic to reach the conclusion is different from mine. You will count the beneficent side of Nature as God's blessing, where I see it as the birthright of our evolutionary origin inside the biosphere. There is no need, however, to stress this conflict between our premises. Instead, let me pose the central part of the naturalistic version, with which I believe you are also likely to agree.

Consider, then, the following truth, which because of its importance deserves to be called the First Principle of Human Ecology: Homo sapiens *is a species confined to an extremely small niche*. True, our minds soar out to the edge of the universe, and contract inward to subatomic particles, the two extremes encompassing thirty powers of ten in space. In this respect our intellects are godlike. But let's face it, our bodies stay trapped inside a proportionately microscopic bubble of physical constraints. We have learned how to occupy some of Earth's most hostile environments—but only when enclosed within airtight containers whose environment is precisely controlled. Polar ice caps, the deep sea, and the moon are ours to visit, but even slight malfunctions of the life-support capsule in which we travel can be terminal to frail little *Homo sapiens*. Prolonged residence there, even when physically possible, is psychologically unbearable.

Here is my point: Earth provides a self-regulating bubble that sustains us indefinitely without any thought or contrivance on our own. This protective shield is the biosphere, the totality of all life, creator of all air, cleanser of all water, manager of all soil, but itself a fragile membrane that barely clings to the face of the planet. Upon its delicate health we depend for every moment of our lives. Humanity, as Darwin observed at the close of *The Descent of Man,* bears the indelible stamp of our lowly origin from preexisting life forms. But even if you cannot agree with that statement for reasons of faith, surely you must grant that we belong in the biosphere, we were born here as a species, we are closely suited to its exacting conditions—and not all conditions, either, but just those in a few of the climatic regimes that exist upon some of the land.

The First Principle of Human Ecology can be put another way: *Alien planets are not in our genes.* If organisms exist on Mars, Europa, or Titanis, then these planets are in *their* genes, and those will surely differ radically from ours.

It follows that human self-interest is best served by not overly harming the other life forms on Earth that still survive. Environmental damage can be defined as any change that alters our surroundings in a direction contrary to humanity's inborn physical and emotional needs. We are not evolving autonomously into something new. Nor are we likely in the foreseeable future to change our basic nature by genetic engineering, as some giddily futuristic writers have envisioned. Scientific knowledge may continue to grow without limit, or it may not. But either way, human biology and emotions will stay the same far into the future, because our immensely complicated cerebral cortex can tolerate little tinkering, because human beings cannot mutate like bacteria to fit every environment we spoil, and because, ultimately, finally and quite simply, we may choose to remain true to human nature, the heritage bequeathed us by millions of years of residence in the biosphere.

Here, then, is another argument for existential conservatism. Beyond the curing of obvious hereditary diseases such as multiple sclerosis and sickle-cell anemia, by gene substitution, the human genome will be modified only at risk. It is far better to work with human nature as it is, by changing our social institutions and moral precepts to get a more nearly optimal fit to our genes, than it would be to tinker with something that took eons of trial and error to create.

The problems of modern civilization rise from the disjunction between our ancient and glacially slow-evolving *genetic* heritage at one level of evolution and our ultrafast *cultural* evolution at the other level. There are still thinkers around the world, some in commanding political and religious positions, who wish to base moral law on the sacred scripture of Iron Age desert kingdoms while using high technology to conduct tribal wars—of course with the presumed blessing of their respective tribal gods. The increasing contrast of such retrograde thinking on the one hand and awesome destructive power on the other should make us more circumspect than ever, and not just about starting wars. It should also make us more careful with the environment, upon which our lives ultimately depend. It will be prudent to curtail the final and permanent obliteration of Nature until we understand more precisely what we are and what we are doing.

The destructive power of *Homo sapiens* has no limit, even though our biomass is almost invisibly small. It is mathematically possible to log-stack all the people on Earth into a single block of one cubic mile and lower them out of sight in a remote part of the Grand Canyon. Yet humanity is already the first species in the history of life to become a geophysical force. We have, all by our bipedal, wobbly-headed selves, altered Earth's atmosphere and climate away from the norm. We have spread thousands of toxic chemicals worldwide, appropriated 40 percent of the solar energy available for photosynthesis, converted almost all of the easily arable land, dammed most of the rivers, raised the planet sea level, and now, in a manner likely to get everyone's attention like nothing else before it, we are close to running out of fresh water. A collateral effect of all this frenetic activity is the continuing extinction of wild ecosystems, along with the species that compose them. This also happens to be the only human impact that is irreversible.

With all the troubles that humanity faces, why should we care about the condition of living Nature? What difference will it make if a few or even half of all the species on Earth are exterminated, as projected by scientists for the remainder of this century? Many reasons exist fundamental to the human weal. Unimaginably vast sources of scientific information and biological wealth will be destroyed. Opportunity costs, which will be better understood by our descendants than by ourselves, will be staggering. Gone forever will be undiscovered medi-

Defining structures of the Pacific yew tree of North America, source of
the anticancer agent taxol. (Original from Charles Sprague Sargent, *Silva
of North America*, 10: plate 514 [1896], reproduced in Eric Chivian, ed.,
Biodiversity: Its Importance to Human Health [Harvard Medical School,
Center for Health and the Global Environment, 2002], p. 19.)

cines, crops, timber, fibers, soil-restoring vegetation, petroleum
substitutes, and other products and amenities.

Critics of environmentalism (whatever that overused term
means—aren't we all environmentalists?) usually wave aside
the small and unfamiliar, which they tend to classify into two

categories, bugs and weeds. It is easy for them to overlook the fact that these creatures make up most of the organisms and species on Earth. They forget, if they ever knew, how the voracious caterpillars of an obscure moth from the American tropics saved Australia's pastureland from the overgrowth of cactus; how a Madagascar "weed," the rosy periwinkle, provided the alkaloids that cure most cases of Hodgkin's disease and acute childhood leukemia; how another substance from an obscure Norwegian fungus made possible the organ transplant industry; how a chemical from the saliva of leeches yielded a solvent that prevents blood clots during and after surgery; and so on through the pharmacopoeia that has stretched from the herbal medicines of Stone Age shamans to the magic-bullet cures of present-day biomedical science.

Because wild natural ecosystems are in plain sight, it is also easy to take for granted the environmental services they provide humanity. Wild species enrich the soil, cleanse the water, pollinate most of the flowering plants. They create the very air we breathe. Without these amenities, the remainder of human history would be nasty and brief. The sustaining matrix of our existence is the green plants, along with legions of microorganisms and tiny invertebrates. These organisms support the world because they are so genetically diverse, allowing them to divide roles in the ecosystem in a fine degree of resolution, and so abundant that at least a few occupy virtually every square meter of Earth's surface. Their functions in the ecosystem are redundant: if one species is eliminated, there is often another able to expand and at least partially take its place. All together the other species, mostly bugs and weeds, run the world exactly as we should want it run, because during prehistory humanity evolved to depend upon their combined actions and the insurance that biodiversity provides world stability.

Living nature is nothing more than the commonality of organisms in the wild state and the physical and chemical equilibrium their species generate through interaction with one another. But it is also nothing less than that commonality and equilibrium. The power of living Nature lies in sustainability through complexity. Destabilize it by degrading it to a simpler state, as we seem bent on doing, and the result could be catastrophic. The organisms most affected are likely to be the largest and most complex, including human beings.

More respect is due the little things that run the world. Being an entomologist, I will now use insects to plead the class-action case on behalf of Earth's entire afflicted fauna and flora. The diversity of insects is the greatest documented among all organisms: the total number of species classified in 2006 is about 900,000. The true number, combining those both known and remaining to be discovered, may exceed 10 million. The biomass of insects is immense: about a million trillion are alive at any given moment. Ants alone, of which there may be 10 thousand trillion, weigh roughly as much as all 6.5 billion human beings. While these estimates are still shaky (to put the matter generously), there is no doubt that insects rank near the top among animals in physical bulk. They are rivaled there in biomass by copepods (minute sea crustaceans), mites (tiny spiderlike arthropods), and, at the very apex, the amazing nematode worms, whose vast population swarms, probably representing millions of species, make up four-fifths of all animals on Earth. Can anyone believe that these little creatures are just there to fill space?

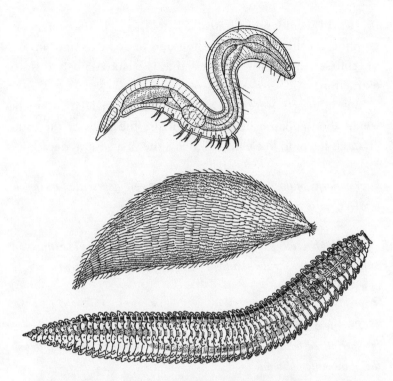

Three species of nematodes (roundworms), specialized variously for
a free-living or parasitic existence. (From Richard C. Brusca and
Gary J. Brusca, *Invertebrates* [Sunderland, Mass.: Sinauer Associates,
1990], p. 350.)

People need insects to survive, but insects do not need us. If
all humankind were to disappear tomorrow, it is unlikely that
a single insect species would go extinct, except three forms of
human body and head lice. Even then there would remain
gorilla lice, closely related to the human parasites and available to carry on at least something close to the ancient line. In
two or three centuries, with humans gone, the ecosystems of

the world would regenerate back to the rich state of near-equilibrium that existed ten thousand or so years ago, minus of course the many species that we have pushed into extinction.

But if insects were to vanish, the terrestrial environment would soon collapse into chaos. Picture the steps of the cataclysm as it would likely unfold across the first several decades:

A majority of the flowering plants, upon being deprived of their pollinators, cease to reproduce.

Most herbaceous plant species among them spiral down to extinction. Insect-pollinated shrubs and trees hang on for a few more years, in rare cases up to centuries.

The great majority of birds and other land vertebrates, now denied the specialized foliage, fruits, and insect prey on which they feed, follow the plants into oblivion.

The soil remains largely unturned, accelerating plant decline, because insects, not earthworms as generally supposed, are the principal turners and renewers of the soil.

Populations of fungi and bacteria explode and remain at a peak over a few years while metabolizing the dead plant and animal material that piles up.

Wind-pollinated grasses and a handful of fern and conifer species spread over much of the deforested terrain, then decline to some extent as the soil deteriorates.

The human species survives, able to fall back on wind-pollinated grains and marine fishing. But amid widespread starvation during the first several decades, human populations plunge to a small fraction of their former level. The wars for control of the dwindling resources, the suffering, and the tumultuous decline to dark-age barbarism would be unprecedented in human history.

Clinging to survival in a devastated world, and trapped in an ecological dark age, the survivors would offer prayers for the return of weeds and bugs.

The bottom line of my scenario is this: be careful with pesticides. Do not give thought to diminishing the insect world. It would be a serious mistake to let even one species out of the millions on Earth go extinct. That is, let me add quickly, with an extremely few exceptions. I'd vote for the eradication of the aforementioned lice (the gravamen against them: limited to humans, serious skin pests, threats to quality of life, carriers of disease). Also, I'd not mourn the passing of mosquitoes of the *Anopheles gambiae* complex of Africa, species that are specialized to feed on human blood, during which they transmit malignant malaria. Keep their DNA for future research and let them go. Let us not be conservation absolutists when it comes to creatures specialized to feed on human beings.

In the real world there is a need to control only the tiny fraction of insect species, perhaps as few as one out of ten thousand, that are consistently harmful to humans. In most cases control means to reduce and if possible to eradicate populations of such species in countries where they are aliens, usually having been transported there by humans as unintended hitchhikers. Take, for example, the red imported fire ant that has vexed the southern United States since the 1940s and has recently spread from there to California, the Caribbean islands, Australia, New Zealand, and China. It inflicts hundreds of millions of dollars in agricultural losses each year. Its stings are painful and occasionally fatal, usually as a result of anaphylactic shock triggered by the venom. It has displaced some native insects and reduced wildlife populations. Obviously it would be wise to erase invading populations of the red fire ant—if only entomologists could find a way. But the same is not true for southern Brazil and northern Argentina, where the ant is not imported but a native species, ecologically adjusted by millions of years of coevolution with other native species. In their South American home they are in balance with predators, pathogens, and competitors. Otherwise they would have become extinct ages ago. In the United States their enemies are fewer in number and weaker. Removal of the alien fire ant populations would be healthy for both people and the environment of the countries they have colonized. Removal from South America, in contrast, might cause damage to the ecosystems in which they are coadapted with other species and live harmoniously.

One of the daunting challenges of the modern discipline of ecology is to sort out such pluses and minuses of living Nature in order better to define the inner structure of the biosphere. There is hope that in time researchers will learn how ecosystems are assembled, how they are sustained, and more precisely how they come to be destabilized. Earth is a laboratory wherein Nature (God, if you prefer, Pastor) has laid before us the results of countless experiments. She speaks to us; now let us listen.

5

Alien Invaders from Planet Earth

ALL RESIDENTS of the southern United States are familiar with fire ants, though on a personal basis. Even though they can be irritating, they also happen to teach us a great deal about how the living world works, or fails to work, and they have become part of American folklore. I knew fire ants intimately myself during my outdoor excursions when I was a boy, and I've studied them off and on throughout my career as a scientist. No other insect better illustrates the delicate complexity of ecosystems, and how easily the balance of nature can be undone by the intrusion of even one alien species. After writing many scientific reports, I thought I was finished with these little stinging demons. I felt I had little more to learn about them. Then a remarkable event brought them back into my life.

I was in the midst of a study of the ants of the West Indies, focusing on them island by island, from Grenada in the extreme south to Cuba and the Bahamas in the north. The whole archipelago is ideal for the study of how plants and animals disperse across water, colonize the land, and form ecosystems. And how they suffer extinction. The 476 kinds of ants occurring on the islands (the latest count, in 2005), by virtue of their abundance and ubiquity, make excellent subjects for such an ecological analysis. The fire ants, as it turns out, were to loom large in the importance of the study for human affairs.

Here is my story.

On the afternoon of March 10, 2003, accompanied by a small group of other field biologists, I entered the excavated ruins of the ancient town of Concepción de la Vega, located in the uplands just west of the center of the Dominican Republic. Straight ahead lay the crumbling stone fort built in 1496 under the direction of Columbus himself. To the left were the remains of an old well, said to have been used by the Franciscan monks who settled here as the sixteenth century dawned. And to the right was a flat area that might well have been part of the monastery garden before this place and the gold rush town around it were abandoned in the 1530s.

A lone sunflower grew on the barren surface of the open space. It was swarming with small dark brown ants. Huddled in the axils of its leaves were families of treehoppers, strange distant relatives of aphids with shark fin spines that protruded from their backs. When I pulled the leaves apart to collect specimens, the ants swarmed over my hands, biting and stinging them. Each sting burned like a match held too close, and most raised a small welt that itched for hours afterward. It was obvious that the ants were protecting the treehoppers.

At that precise moment, in those odd circumstances, I felt confident I had solved a 500-year-old mystery. At last, as the culmination of considerable effort, I could report the cause of the first environmental crisis experienced by European colonists of the New World.

Around 1518, a plague of ants irrupted at the fledgling Span-

ish colony on Hispaniola. The event was witnessed by Fray Bartolomé de Las Casas, exacting chronicler of Columbian America ("who promises before the divine word that everything said
and referred to is the truth") and defender of the Caribbean
Indians. A great saint, in my opinion, never canonized. He
described the scene at the monastery as follows in his *History of
the Indies*. "This plague was an infinite number of ants that . . .
bit and caused greater pain than wasps that bite and hurt men.
They could not defend themselves from these ants at night in
their beds, nor could they survive if the beds were not placed on
four small troughs filled with water."

Elsewhere, in the newly established capital of Santo Domingo
and in other parts of what is today the Dominican Republic, ant
swarms destroyed gardens and orchards everywhere. As the
plague spread, entire plantations of oranges, pomegranates,
and cassias were wiped out. "As though fire had fallen from the
sky and burned them," Fray Bartolomé agonized, "they stood
all scorched and dried out." The loss of the cassia trees, source
of a purgative widely used in Spain, was particularly distressing.
The colonists, whose income from mining had dropped with the
near-extinction of the enslaved Taino Indians from maltreatment and disease, had turned to this crop as an important new
source of income.

Fray Bartolomé believed that the plague was an expression of
God's wrath for the maltreatment of the Taino people. Whatever the Spanish themselves thought about the cause, they soon
turned to the highest level of authority for relief:

As the citizens of Santo Domingo saw the affliction of this
plague grow, doing such damage to them, and as they

Fray Bartolomé de Las Casas, historian of Columbian
America (1484–1566). (© Corbis.)

could not end it by any human means, they agreed to ask
for help from the Highest Tribunal. They made great pro-
cessions begging Our Father to free them from such a
plague so harmful to their worldly goods. In order to
receive divine blessing more quickly, they thought of tak-
ing a saint as a lawyer, whichever one by chance our Lord
should declare best suited. Thus, with the procession over

one day, the bishop, and clergy, and the whole city cast lots over which of the litany's saints Divine Providence would see fit to give them as a lawyer. Fortune fell on Saint Saturnin, and receiving him with happiness and joy as their patron, they celebrated him with a feast of great solemnity, as they have each year since then. . . .

And indeed, according to Fray Bartolomé, the plague, as if miraculously, soon began to recede. Within a few years new trees were planted and brought to fruit. To this day citrus and cassia trees flourish throughout the Dominican Republic, and they remain mostly free of damage from ants.

As the ant plague subsided on Hispaniola, it struck elsewhere in the West Indies. Early in the 1500s, an onslaught of the insects contributed to abandonment in 1534 of the village of Sevilla Nueva on Jamaica. About the same time, swarms of ants threatened the cassava plantations on what is today Loíza on Puerto Rico, and after casting lots the people named Saint Patrick their protector. When a similar plague afflicted Sancti Spíritus in Cuba, the population moved across the river, and Saint Ann was selected for intercession.

In the seventeenth century, ants reached near-plague levels on Barbados, an event Richard Ligon related in his firsthand 1673 description of natural history on the island. In the eighteenth century a full-blown plague swept through the Lesser Antilles: Barbados in 1760, Martinique in 1763, and Grenada in 1770. On the last island, R. H. Schomburgk later wrote in his 1848 *History of Barbados,* "every sugar-plantation between St. George's and St. John's, a space of about twelve miles, was destroyed in succession, and the country was reduced to a state of the most deplorable condition." The ants were so dense, he added, that they covered roads for miles at a stretch. Impressions made by the hooves of horses traveling along the road remained visible for only a moment or two, until they were filled up by the ants.

No saints were selected to save the sugarcane crops of the Lesser Antilles, but large rewards were offered—£20,000 in the case of Grenada—to anyone who found a way to stem the formicid tide. None was forthcoming, but in the end it did not matter. On these islands, as on Hispaniola more than two centuries earlier, the plagues diminished on their own.

What was the plague ant? This was a mystery of identity, rather like a criminal investigation. Carolus Linnaeus, founder of modern taxonomic classification, gave the plague ant the Latin name *Formica omnivora* (omnivorous ant) in 1758. But that was all he gave it. Today his sparse Latin diagnosis provides no clear idea of the identity of the species in modern classification systems. Nor have I or other entomologists been able to locate authenticated specimens in the Linnaean collections of Stockholm or London that allow us to place it exactly. Ant experts of the past, including the learned William Morton Wheeler, one of my predecessors as curator of entomology at Harvard, had speculated on which of the ant species still in the Caribbean region was the culprit, but the evidence they had was too thin and contradictory for them to draw a firm conclusion. Wheeler, as it eventually turned out, came close in a 1926 article on the subject, but his guess was not squarely on target. To use a courtroom analogy, investigators from Wheeler forward had suspects but not enough evidence to bring in an indictment.

The enigma of the West Indies plague ants has historical significance (for example, few other creatures have had saints assigned them). But beyond that its resolution is relevant to our general understanding of unstable environments. What was *Formica omnivora*? Why did it explode to plague proportions? And finally, why did it then recede within a few years or at most decades?

In the mid-1990s I set out to see whether I could close this cold case of entomology. From time to time, I visited the islands on which the plagues had occurred, examining all the ant species I could locate in their present-day habitats. Poring over the historical literature, I pieced together every available scrap of information about the appearance and behavior of *Formica omnivora*. From this information I drew a short list, then a shorter list. In the end, after much wavering and several reversals, I made my decision from what I found at the monastery of Concepción de la Vega.

The plague ant of the sixteenth century, I concluded (as had Wheeler on less evidence), was the tropical fire ant. Known to entomologists by its scientific name *Solenopsis geminata,* it is evidently native to the extreme southern United States, Central America, and probably tropical South America, but has been spread by human commerce over a large part of the tropical and subtropical parts of the world. It is a different species from the red imported fire ant of the southern United States. The two fire ant species most closely related to it occur in the southwestern United States. The tropical fire ant may also be native to the West Indies. At least it was present when Columbus first came ashore. The Taino people had a name probably applied to it, *jibijoa,* which they were unlikely to have invented from 1492 to their final extermination at the hands of the Spanish four decades later. If not truly native but still at least pre-Columbian in origin, the ant was accidentally transported by the Taino's Arawak ancestors from island to island up the Lesser Antillean chain of islands. One excellent candidate for the vehicle of transport would have been stocks of cassava, a favored root vegetable of the indigenous Caribbean people.

Yet—here the mystery deepened. If the fire ants lived in and around the Taino gardens, why did this insect wait for Columbus to arrive to irrupt into a plague? Assuming that the outbreak was not God's punishment for the genocide of the Taino (I cannot entirely exclude that hypothesis!), the cause must have been something the Spanish did to the environment. It could not have been simply the planting of orchards and gardens. Hispaniola was already cultivated heavily by the 400,000 or so Taino present before the Spanish occupation.

The solution, I realized upon seeing the ants and treehoppers at Concepción de la Vega, lies in the scorched appearance of the dying crop plants. This is not an effect produced by any known ants, which very seldom consume plant material. But it does result from heavy infestations of sap-sucking homopterous insects, including aphids, mealybugs, scale insects—and treehoppers. Fire ants are among the kinds of ants that protect these insects, and in exchange the homopterans provide them with liquid excrement rich in sugar and amino acids. It appears that the most likely cause of the plagues was the arrival of one or more homopterans new to Hispaniola. These pests, carried to the island inadvertently by the Spanish, and at first unopposed by any parasites or predators natural to them, bloomed into dense populations. The most likely vehicles in this case were plantains, brought from the Canary Islands in 1516 to be cultivated as a major food crop. The ants, profiting from the increased food supply, luxuriated in their newfound pastures, and the symbiosis of the two kinds of insects created the plague.

The Spanish, not noticing the homopterous sapsuckers amid the myriad kinds of insects abounding around their crops, or at least not grasping their significance, understandably put all the blame on the viciously stinging ants. It was not until the late eighteenth century, on Grenada, that naturalists began to suspect the involvement of homopterans in the West Indian ant plagues.

My confidence in the identity of the mysterious *Formica omnivora* of sixteenth-century Hispaniola was rein-

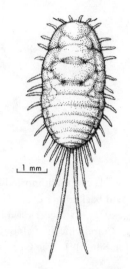

A mealybug *(Pseudococcus longispinus)*. (Courtesy of CSIRO, Department of Entomology. From T. E. Woodward, "Hemiptera," in *The Insects of Australia* [Melbourne: University of Melbourne Press, 1970], p. 429.)

forced by having witnessed a fire ant plague on my own—and from near its beginning. Sometime in the late 1920s or early 1930s, the aforementioned red imported fire ant (scientific name: *Solenopsis invicta*) was accidentally introduced into the port of Mobile, Alabama. It had almost certainly hitchhiked in seaborne cargo from somewhere within its native range in central Brazil and northernmost Argentina, most likely along the shipping lanes of the Paraná River. In 1942, as a thirteen-year-old, I happened to be studying ants for a Boy Scout project in the neighborhood

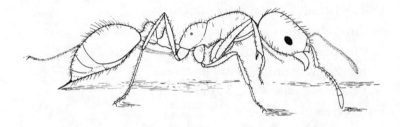

A worker of the imported fire ant lays an odor trail from newly discovered
food back to her colony's nest. The trail pheromone is released from the
extruded sting. (Drawn by E. O. Wilson, from E. O. Wilson, "Chemical
Communication among Workers of the Fire Ant *Solenopsis saevissima*
[Fr. Smith], 1: The Organization of Mass-Foraging," *Animal Behaviour* 10,
no. 1–2 [1962]: 134–47.)

where I lived, only half a dozen city blocks from the Mobile
docks. I discovered a characteristic high mound nest of the red
imported fire ant, one of the earliest two records made of this
species in the United States. Seven years later, the ant had
spread outward in all directions from Mobile for eighty miles,
building dense populations on lawns, fields, and grassy road
strips, reaching fifty or more mound nests per acre, each con-
taining as many as 200,000 ill-tempered worker ants. It can then
be said to have reached plague proportions, perhaps not as
severe as that of the common fire ant in sixteenth-century Hispaniola, but enough to cause widespread hardship and alarm.

In the spring of that year, 1949, I was a senior at the University of Alabama, deep into my studies of entomology and with a
special interest in the biology of ants. The Alabama Department
of Conservation hired me to conduct a survey of the red

imported fire ant and its impact on the environment. Not yet twenty, I had landed my first job as an entomologist! I owed a lot to the fire ant. I really could, I realized, make a living from my boyhood passion. By crisscrossing the infested area, in the company of a fellow undergraduate, Jim Eads, I soon confirmed the ominous earlier reports. The ants I observed in the field, and checked with laboratory experiments, were significantly harming crops, especially in garden plots, by carrying off seeds and eating into the roots of seedlings. I recorded many reports of the insects attacking the pipping chicks of bobwhite and other birds nesting on or close to the ground. I could see how the ants and their large nests often made plowing, mowing, and harvesting difficult. I noted that they sometimes invaded houses, especially in rural areas. All of these infelicities have been confirmed by later researchers. The more recent studies have revealed much more. The fire ants alter the environment by reducing the abundance and diversity of many other insect and other invertebrate species, as well as reptiles, and they are powerful enough even to displace or diminish populations of mice and deer. A small percentage of humans, fortunately fewer than 1 percent, develop an allergic reaction to the venom.

The joke in fire ant country today is that the name of the famous pest is pronounced "far aint." To which the teller adds quickly, "We're not talking southern dialect here; we're saying the ants came from far away and they ain't going back." That turns out to be an understatement. The red imported fire ant is almost unstoppable; it lives up to its Latin scientific name, *invicta*—the unconquered one. Once established, the population spread through the Gulf States, moving northward as well until the freezes of winter proved too much for its warm-climate physiology. It now ranges continuously from the North Carolina flatlands to central Texas, and south through all of Florida. In the 1980s it vaulted over to Puerto Rico, carried by human commerce no doubt, and has since made its entrance into the Bahamas, parts of the Lesser Antilles, and Trinidad. In the 1990s it colonized Orange County, California. As I recently told my entomologist colleagues at the University of California at Davis, in the Central Valley, "First you'll hear a hissing sound off to the south, and then they'll be here."

As it turned out, all of these events added up to only the first chapter of the ant plague epic. As I put all the pieces together in my effort to prove the identity of the West Indian species, I realized the evidence still contained two discrepancies. First, the ants that flooded Barbados, Grenada, and Martinique in the mid-1700s did not sting! Or, at least, no mention of this overwhelmingly obvious trait of fire ants is made in contemporary records. To be stung by fire ants, an unavoidable experience on close contact, is to resolve to mention the experience prominently in future dispatches. And second, Richard Ligon, describing in 1673 a species of plague or near-plague proportions, reported that when the ants found a food parcel too heavy for a single worker to carry (such as dead cockroaches, which he squashed and fed to the ants for amusement), picked it up in a group and carted it back to the nest in unison. Fire ants, in contrast, either drag large food objects or carve them into smaller pieces that can be carried by individuals.

It was thus apparent that two kinds of ants caused the West Indian plagues: fire ants in the sixteenth century on Hispaniola and then something else a century or more later to the south in the smaller islands of the Lesser Antilles. In the latter case, the prime and virtually only suspects left are species of *Pheidole* ants. The ant genus *Pheidole* is the most diverse and abundant of the Western Hemisphere, with some 624 species known to science. Because I had just completed a thorough study of all of them, including describing 344 species new to science, I knew at once that two candidates were possible: Jelski's Pheidole (scientific name: *Pheidole jelskii*) and the Big-Headed Ant Pheidole *(Pheidole megacephala)*.

I could eliminate *Pheidole jelskii* quickly. Although this native species is one of the most abundant and widespread ant species of the New World, and occurs throughout the West Indies, it does not otherwise fit the historical profile of a plague species. It makes crater-shaped nests in open fields, it does not invade houses, and it does not collect in large masses. *Pheidole megacephala,* on the other hand, fits the profile almost perfectly. An alien species of African origin, it nests in the roots of trees and sugarcane, in the manner recorded for the plague ants, and it is often a major house pest, as reported in Ligon's seventeenth-century accounts. Further, it forms gigantic, continuous colonies able to dominate isolated areas completely. I found such a supercolony on Loggerhead Key in the Dry Tortugas of Florida, and other entomologists have reported them on Bermuda and on Culebrita, near Puerto Rico. In a few other parts of the world, including Hawaii, this ant has reached plague or near-plague proportions in recent years.

If the alien *Pheidole megacephala* was indeed the second plague ant, another peculiar feature of the West Indian players makes sense. The three worst outbreaks after the 1500s, those on Barbados, Grenada, and Martinique, all commenced during 1760–70, in other words during a ten-year period, and all were concentrated in fields of sugarcane. It is hard to explain this phenomenon except as the relative late arrival of an alien— either *Pheidole megacephala* itself or, more likely, since a plague ant was already present on Barbados in the mid-1600s, newly introduced homopterous insects with which it formed a symbiosis. The latter explanation is strengthened by the fact that the plague was focused in the cane fields, where homopterans can multiply in great numbers.

Of the slightly fewer than twelve thousand ant species known in the world, just thirteen have become invasive by hitchhiking on human commerce, colonizing new areas, then causing significant amounts of ecological or economic damage. Most have moved on up to plague proportions at one time or another. In addition to the fire ants and the Big-Headed Ant, this select group includes the secretive ninja ant *(Monomorium destructor)* on Cape Verde and the little fire ant *(Wasmannia auropunctata)*, which is devastating insects and other small animals in the Galápagos Islands, New Caledonia, and other tropical localities. The Argentine ant *(Linepithema humile)*, another cosmopolitan ant pest, has been the scourge of Madeira and parts of Australia, South Africa, and California.

It is not surprising that such tiny insects could have such a big impact. Ants are, after all, among the dominant small animals of the planet. In the Amazon forest, where measurements of the phenomenon have been made, they compose one-third of the insect dry weight, and together with termites more than a quarter of the dry weight of all animals, vertebrate and invertebrate, combined. Those figures are likely equaled or approached elsewhere, at least in savannas, deserts, and even warm temperate forests. Ants move more earth than earthworms and are the

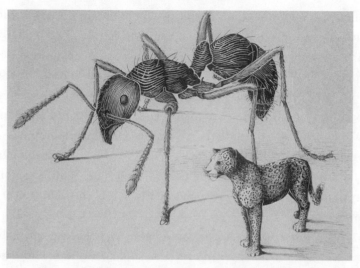

In the Brazilian Amazon, and probably many other habitats as well, ants outweigh all land vertebrates together (mammals, birds, reptiles, amphibians) four to one. (Drawing by Katherine Brown-Wing, in E. O. Wilson, *Success and Dominance in Ecosystems* [Oldendorf/Luhe, Germany: Ecology Institute, 1990], p. 5.)

principal predators and scavengers of small animals in most habitats. It is questionable whether humanity could survive without them, even if other insects survived. Their ecological regnancy makes ants more likely than other animals to be transported by humans. For every harmful species, moreover, there are at least ten aliens established in some part of the world or other (many, especially, in the southeastern United States) that are not pests—at least not yet.

The ant story is a fateful reflection of what is happening to the rest of life on the planet. As globalization and international commerce and travel increase, so does the rate of spread of alien species entirely as a result of human activity. Every country is the largely unconscious host of a rising tide of such aliens. The number of immigrant plant, animal, and microorganism species listed by the federal government in 1993 was 4,500, compared with a total of some 200,000 known native species. But this is surely an underestimate. The true number of invaders, when rare and still hidden species of small invertebrates and microbes are added, could easily be in the tens of thousands. In Hawaii, the most biologically transformed state of all, a majority of the resident land birds and nearly half of the plant species are alien.

The United States has been invaded by alien species throughout its history. When agricultural pests and exotic human disease agents are included in the roster, the total cost runs into the hundreds of billions of dollars annually. The damage is of multiple kinds. An Asian fungus, for example, wiped out the American chestnut, the dominant tree of the eastern American forests in the early 1900s. Zebra mussels from either the Black Sea or the Caspian Sea, spreading from their American point of introduction in the Great Lakes, now clog the intake valves of electrical utilities and alter freshwater ecosystems. But the species that most chills my conservationist heart is the brown tree snake from the southwestern Pacific. In a few decades following its introduction onto Guam after World War II, it erased completely all ten of the native forest bird species living on the island, including three found nowhere else in the world. And as if this were not bad enough, the snakes are also poisonous, grow to ten feet, and occasionally enter houses.

These invaders are only part of the spearhead. Among other recent immigrants that live comfortably in the United States are tiger mosquitoes, Formosan termites ("the termites that ate New Orleans"), pond-hopping snakehead fishes, miconia (a "green cancer" on shrubs and trees), and the balsam tree adelgid, an aphidlike homopteran that has destroyed a large part of the southern Appalachian fir forest. I have taken perverse pleasure in stringing together the titles of five recent (and very good) books detailing the impact of the invasive species, in order to tell the story in one sentence: *Alien Invaders* are a form of *Biological Pollution*; as *Strangers in Paradise* and *Life Out of Bounds,* they have become *America's Least Wanted.*

Around the world, invasive species are the second-ranking cause of extinction of native species, after the destruction of habitats by human activity. In the long term they are slowly changing the biological quality of our planet. Because we have had only limited success in controlling them, we are left in most cases with no choice except to wait them out, as the people of the Caribbean Islands did the tropical fire ant and its presumed symbiont. Given enough time, most settle down to live within or at least alongside what is left of the ecosystems they have threatened.

The reasons for the subsidence of the invasives remain largely unknown. They are likely to include the increase in numbers and effectiveness of the parasites, predators, and competing species able to adjust to them. How long does the process take? No record was kept by the early chroniclers, but the plague ants of the West Indies apparently took from several years to several decades to fall back to at least seminormal levels. After sixty years the red imported fire ant appears to be subsiding somewhat in the southern United States; in this case strenuous efforts at control have had at least a local effect.

In the long term the most insidious impact of the rising alien tide is the homogenization of Earth's ecosystems. As native species retreat and disappear, to be replaced by superior alien competitors from other lands, global biodiversity is declining and with it the differences in life forms from one place to another. The bird with the orange-red head flashing through the almost wholly alien rainforest of lowland Oahu is the same one you'll see in southern Florida and its native home of Brazil. The beautiful purple loosestrife that graces the swampy meadows of North America, where it crowds out native plants, is the same species that ranges from its homeland in Europe to Japan and thence to outposts in Ethiopia, Australia, and New Zealand.

The homogenization of the biosphere is painful and costly to our own species and will become more so. If we are to stem it, we will have to learn more about biodiversity and what is happening to that most precious natural resource. Let us think upon what we and the other aliens are doing to the rest of life, and to ourselves.

6

Two Magnificent Animals

PASTOR! No words and no art can capture the full depth and intricacy of the living world—as biologists have come increasingly to understand it. If a miracle is a phenomenon we cannot understand, then all species are something of a miracle. Each and every kind of organism, by virtue of the exacting conditions that produced it, is profoundly unique and shows its diagnostic traits reluctantly.

To press this key point, let me tell you about two of the species I personally have found of compelling interest.

THE WOLVERINE

I have never seen a wild wolverine, and I hope I never will. This weasel-like mammal of the north woods is legendary for its ferocity, cunning, and elusiveness. Chunky in form, three to four feet long and weighing twenty to forty pounds, it is one of Earth's smallest top-tier predators. It feeds on everything from rats to deer. It can chase cougars and wolf packs away from downed prey, and drag carcasses three times its own weight. It has fuzzy thick black fur, but this is no animal you'd want to pet. It has sharp teeth, a predator's retractable claws, and the face of a miniature bear. It walks flat-footed and low to the ground, such that when standing still it seems poised to spring forward. The American naturalist Ernest Thompson Seton wrote of this species in 1908:

> Picture a weasel—and most of us can do that, for we have met that little demon of destruction, that small atom of insensitive courage, that symbol of slaughter, sleeplessness, and tireless, incredible activity—picture that scrap of demoniac fury, multiply that mite fifty times, and you have the likeness of a wolverine.

The other vernacular names given it, devil bear, skunk bear, carcajou, and glutton, and even its brutish scientific name, *Gulo gulo,* suggest the gap that exists between the wolverine and humanity. Add to that the difficulty of spotting a wolverine in

A wolverine. (Plate from *A Field Guide to Mammals of Britain and Europe* by F. H. van den Brink, translated by Hans Kruuk and H. N. Southern, Illustrated by Paul Parruel [Boston: Houghton Mifflin, 1968].)

the wild. Individuals are both solitary and exceptionally shy of humans. They wander far and wide—here today, over there somewhere tomorrow, and gone for good the day after that.

Its savage demeanor is not, however, why I want to avoid the wolverine. The reason is that I find *Gulo gulo* the embodiment of wildness, and I know there will still be untrammeled habitats on Earth if wolverines still roam there. I trust they will hold on in the vast subarctic forest, somewhere in North America or Eurasia, in places too far to be reached easily by vehicles or shank's mare. Wildlife biologists will need to know the general status of the wolverine in order to save the species, but I hope there always will be remote sections of its range barred to trappers and even scientists. Please let part of the wolverine stay a mystery!

One day while I was on a visit to the University of Montana, in Missoula, a professor of biology told me the kind of story I most like to hear. A neighbor, he said, was running camera traps in his backyard. The house was very near the border of the Rattlesnake Wilderness, which continues on from Missoula through the forested corridor of the northern Rockies. A camera trap takes photographs of animals that touch a trip wire or break an electronic beam. It is especially good at catching images of rare and shy nocturnal animals otherwise never seen. Among the prints from a few nights' recent coverage, my friend said, was— imagine this—a *wolverine*! A few are known to wander into the lower forty-eight from their fastnesses in the Canadian provinces. The chances of ever seeing one in the flesh are minimal, yet there is a thrill in just knowing that they exist, and that they might even pay a brief, unseen visit to the neighborhood.

The incident illustrates what has been called the Grizzly Bear Effect of environmental ethics. We may never personally glimpse certain rare animals—wolves, ivory-billed woodpeckers, pandas, gorillas, giant squid, great white sharks, and grizzlies come to mind—but we need them as symbols. They proclaim the mystery of the world. They are jewels in the crown of the Creation. Just to know they are out there alive and well is important to the spirit, to the wholeness of our lives. If they live, then Nature lives. Surely our world will be secure, and we will be better for it. Imagine the shock of the following headline: LAST TIGER SHOT, SPECIES NOW SAID EXTINCT.

THE PITCHFORK ANT

This is how I see living species: masterpieces, legends. I hope I live long enough to see the Grizzly Bear Effect extended to a few small creatures as well. Such a pleasurable focus is, I grant, an acquired taste. But it can also carry an emotional wallop. My favorite examples are ants belonging to the genus *Thaumatomyrmex*. This scientific name, drawn from the Greek, means "wonderful ant." The genus comprises a dozen species distributed across different parts of the New World tropics. They are the rarest ants in the world, or close to it. In my entire career, during repeated trips to places where *Thaumatomyrmex* might be found, I have collected exactly two specimens. The catch of even one individual—or still more dramatically an entire colony—is a newsworthy event among specialists on the biology of ants, admittedly not a large group of people.

Thaumatomyrmex are not stereotypical ants, the familiar kind that swarm in and out along odor trails from earthen nests in woods and fields. *Thaumatomyrmex* colonies are tiny, comprising

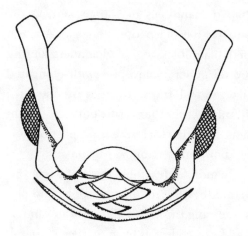

The head of *Thaumatomyrmex paludis,* from Isla Tórtula, Venezuela. (From Neal Weber, "The Genus *Thaumatomyrmex* Mayr with Description of a Venezuelan Species [Hym.: Formicidae]," *Boletin de Entomologia Venezolana* 1, no. 3 [1942]: 65–71.)

at most ten or twenty members that hide in unformed nests within pieces of decaying wood on the tropical forest floors. Foragers go forth to hunt food in solitude. They do not follow trails, and they carry their prey home singly without help from nestmates.

The fame of *Thaumatomyrmex* among the ant cognoscenti arises not from their scarcity, however, but from their bizarre anatomy. Their heads are completely unlike those of any other of the known species of ants: short, concave in front, and bearing enormous jaws shaped like pitchforks. The teeth, or tines of the pitchfork, are sometimes so elongate that when the mandibles are closed, the largest pair curve all around the opposite side of the head and stick out behind its posterior rim. What is the function of those strange instruments? That is indeed the question of interest. Myrmecologists have studied many kinds of ants with oddly shaped jaws, and their idiosyncrasies always turn out to serve some highly specialized purpose. Army ant soldiers employ their sickle-shaped mandibles in battle, piercing the skins of opponents with needle-sharp tips. *Polyergus* Amazon ants wield saber-shaped mandibles to kill defenders during slave raids. Ants in several genera snare prey with elongated jaws that snap shut like animal traps. In at least one such species, the teeth would travel faster than a rifle bullet, if the ant were the size of a human being. Their closure is proportionately the quickest movement known in the animal world.

The *Thaumatomyrmex* mandibles resemble none of these various forms. What, then, are they for? In an attempt to find out, I once spent four days tramping through a Costa Rican rainforest where specimens had been collected by an earlier entomologist. Frustrated and depressed, I found not a single worker. I then published an appeal in *Notes from Underground,* the ant biologists' newsletter. There are, I wrote, several things I want most to know about ants before going up to that Great Rainforest in the Sky. One of the mysteries I needed to have solved for peace of mind is what *Thaumatomyrmex* does with its pitchfork mandibles.

The appeal worked. There is nothing more satisfying to younger scientists than showing up older ones. Not long afterward two (young) Brazilian entomologists spotted a *Thaumatomyrmex* carrying a prey item. They were able to track it back all the way to the nest. Here is what they found—and was later confirmed at a locality in the Amazon by a German entomologist. *Thaumatomyrmex* is a specialized predator on polyxenid millipedes. Most millipedes, which are popularly known as "thousand legs," are covered by hard chitinous armor plates that shield them from attacks by ants and other enemies. Polyxenids have soft skins and are protected instead by a dense coat of long bristles. They are the porcupines of the millipede world. The *Thaumatomyrmex* foragers are porcupine huntresses. They slip the tines of their pitchfork jaws through the bristles, pierce the bodies of the polyxenids, and carry them home. There they scrape off the bristles with specialized brushes on their forefeet, rather like farmers plucking chickens. Afterward they carve up the millipede, and share the morsels with nestmates.

THE GREATEST HERITAGE

For professional and serious amateur naturalists alike, there are countless wonders like those inherent in wolverines and the *Thaumatomyrmex* ants. They range in scientific importance from minor to paradigm breaking, and in organisms from bacteria to whales and from algae to redwoods. For those who love adventure and real-world challenges, body and mind, Nature is a heaven on Earth. Here, Pastor, we surely agree. The Creation, whether you believe it was placed on this planet by a single act of God or accept the scientific evidence that it evolved autonomously during billions of years, is the greatest heritage, other than the reasoning mind itself, ever provided to humanity.

7

Wild Nature and Human Nature

OUR RELATIONSHIP to Nature is primal. The emotions it evokes arose during the forgotten prehistory of mankind, and hence are deep and shadowed. Like childhood experiences lost from conscious memory, they are commonly felt but rarely articulated. Poets, at the highest human level of expression, try. They know that something fundamental moves beneath the surface of our conscious minds, something worth saving. It evokes some of the spirituality that you and I, Pastor, hold in common.

Thus has been born a different kind of literature, and with it an impulse toward the conservation of Nature. George Catlin, the premier portraitist of the American Indian, expressed the creative impetus very well in his notes of 1841:

Many are the rudenesses and wilds in Nature's works, which are destined to fall before the deadly axe and desolating hands of cultivating man; and so amongst her ranks of the *living*, of beast and human, we often find noble stamps, or beautiful colours, to which our admiration clings; and even in the overwhelming march of civilized improvements and refinements do we love to cherish their existence, and lend our efforts to preserve them in their primitive rudeness.

The gravitational pull of Nature on the human psyche can be expressed in a single, more contemporary expression, *biophilia,* which I defined in 1984 as the innate tendency to affiliate with life and lifelike processes. From infancy to old age, people everywhere are attracted to other species. Novelty and diversity of life are esteemed. Nowadays the word "extraterrestrial" summons in ultimate manner the countless images of still unexplored life, replacing the old and once potent "exotic," which drew earlier travelers to unnamed islands and remote jungles. To explore and affiliate with life, to turn living creatures into emotion-laden metaphors, and to install them in mythology and religion—these are the easily recognized fundamental processes of biophilic cultural evolution. The affiliation has a moral consequence: the more we come to understand other life forms, the more our learning expands to include their vast diversity, and the greater the value we will place on them and, inevitably, on ourselves.

Two new academic disciplines have emerged that address the twin subjects of biophilia and conservation in a systematic manner. Environmental psychology covers all aspects of the relation of human mental development to the environment. Conservation psychology in turn focuses on the many facets of biophilia in order to help design the most effective conservation procedures for natural environments and species.

It is in human mental development that the perceptions of living Nature and human nature unite, as well as science and the religious experience. Our connections to the rest of life, and to all the love, art, myth, and destructiveness that flow into culture from that relationship, are products of the interaction of instinct and environment. The instinctual part is what we call human nature.

What precisely, then, is human nature? That is one of the great questions of both science and philosophy. It is not the genes that prescribe human nature. It is not the cultural universals, such as incest taboos, rites of passage, and creation myths. Those are the products of human nature. Rather, human nature is the hereditary rules of mental development. The rules are expressed in the molecular pathways that create cells and tissue, particularly those of the sensory and nervous systems. The rules are also prescribed in the cells and tissues that generate mind and behavior. They are manifested as biases in the way our senses perceive the world. They appear as the properties of language and symbolic coding by which we represent the world. The developmental rules are not absolute. Instead, they generate the options we open to ourselves. They render some choices more pleasing than others: music yes, the crying of a baby no.

The developmental rules are in an early stage of exploration by psychologists and biologists. Even so, the few that are known range over diverse categories of behavior and culture. They affect how we clarify colors in accordance with the innate coding of cell reception and transmission within the retina. They bias our aesthetic response to visual design according to elementary abstract shape and degree of complexity.

In a wholly different realm, developmental rules determine the readiness by which we acquire aversions and phobias. People come most quickly to fear objects that were dangerous to prehistoric people, including snakes, spiders, heights, closed space, and other ancient perils of humankind. The trigger that creates one of the deep aversions is often a single frightening experience. To be startled by a sudden writhing of an object on

The terror and power of the serpent is expressed in most human cultures.
This is a depiction of a combined Andean snake-cat personage, probably
the hero Ai-Apaec. (From Balaji Mundkur, *The Cult of the Serpent* [Albany:
State University of New York Press, 1983], p. 129.)

the ground can imprint the mind against snakes. I escaped that
phobia somehow. In fact, I have always enjoyed catching and
handling snakes, a taste learned as a boy naturalist. On the
other hand, I have a mild and unshakable arachnophobia,
acquired during an accidental entanglement with the web of a
large orb-weaving spider when I was eight years old. I enjoy
exploring caves—no claustrophobia there—but because of a
clumsy anesthesia during an operation when I was a small boy,
I cringe at even the thought of my face being covered while my
arms are pinned. In general terms, I'm typical. Every person
has his own imprinting experience and profile of such archaic
aversions. Only a lucky few lack them completely.

In sharp contrast to their inborn sensitivity to ancient perils,
people are far less prone to acquire fear of knives, guns, auto-
mobiles, electric outlets, and other dangerous objects of every-
day modern life. The reason for the difference, scientists
believe, is insufficient time for the evolving species to hardwire
reactions in the brain to these newer threats.

And what of biophilia? A good example lies in plain sight. Researchers have found that when people of different cultures, including those of North America, Europe, Asia, and Africa, are given freedom to select the setting of their homes and work places, they prefer an environment that combines three features. They wish to live on a height looking down and out, to scan a parkland with scattered trees and copses spread before them, closer in appearance to a savanna than to either a grassland or a closed forest, and to be near a body of water, such as a lake, river, or sea. Even if all these elements are purely aesthetic and not functional, as in vacation homes, people who have the means will pay a very high price to obtain them.

There is more. Subjects in choice tests prefer their habitation to be a retreat, with a wall, cliff, or something else solid to the rear. They want a view of fruitful terrain in front of the retreat. They like large animals scattered thereabout, either wild or domestic. Finally, they favor trees with low horizontal branches and divided leaves. It is probably not a coincidence that some people, I among them, consider the Japanese maple the world's most beautiful tree.

These quirks of human nature do not prove but are at least consistent with the savanna hypothesis of human evolution. Supported by considerable evidence from the fossil record, this interpretation holds that human beings today still choose the habitats resembling those in which our species evolved in Africa during millions of years of prehistory. The distant forebears wished to be hidden in copses looking out over a savanna or transitional woodland, scanning the terrain for prey to stalk, fallen animals to scavenge, edible plants to gather, and enemies to avoid. A body of water nearby served as a territorial boundary and an added source of food.

By and large, people are keenly aware of their own innate

A Japanese maple *(Acer palmatum).* (Photo by Peter Gregory. From J. D. Vertrees, *Japanese Maples: Momiji and Kaede,* 3rd ed. [Portland, Oreg.: Timber Press, 2001], p. 67. Used by permission.)

preferences but have given little or no thought to why they and others feel the same way. I once dined at the home of the late Gerard Piel, a distinguished writer, publisher, and the founder of *Scientific American.* He was, I knew, disinclined to accept the idea of a genetic human nature. So it gave me considerable pleasure to stroll out with him onto the balcony of his penthouse apartment, which was lined with potted shrubs, and gaze down with him more than a dozen stories to the woodland, savanna, and reservoir lake of Central Park. I can only imagine how much that view added to the commercial value of the apartment— thanks to choices made by our long-ago African ancestors.

Is it so strange that at least a residue of habitat selection persists among the human instincts? The programmed search for the correct environment is a universal of animal species for the best of

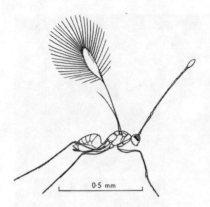

0·5 mm

A fairy fly (species of *Mymar*). (Courtesy of CSIRO, Department of Entomology. From E. F. Riek, "Hymenoptera," in *The Insects of Australia* [Melbourne: University of Melbourne Press, 1970], p. 916.)

reasons—it is an imperative of survival and reproduction. My favorite example, being an entomologist, I suppose, is the behavior of the fairyfly, a tiny myrmarid wasp that parasitizes the eggs of dytiscid water beetles. After flying about and finding the right places to mate, the female begins her search for prey. She lands on the surface of an appropriate body of water that may hold eggs. She stands there at first, her little body held in place by surface tension. To submerge she digs—she is too light to dive—through the surface tension with her legs. She then swims downward, using her wings as paddles. Reaching the bottom, she searches about like a pearl diver for eggs of the water beetles, into which she inserts her own eggs. All this is accomplished with a brain no larger than a dot made by a fine pen.

Returning to *Homo sapiens,* it would be quite extraordinary to find that all the rules of learning bias related to the ancestral world have been erased during the past several thousand years. The human brain is not and never was a blank slate.

Given that the natural world is still embedded in our genes and cannot be eradicated, we should see its effect not only on our habitat preference but on other aspects of our mental and physical well-being. Psychologists have in fact discovered that just a view of natural environments, especially parklands and savannas, generally leads to a decline in moods of fear and anger, and it generates an overall feeling of tranquility. In one study, postsurgical patients who were allowed to look out at trees, recovered more quickly and reported less need for pain and anxiety medication than those treated otherwise identically but whose view was the walls of buildings. In parallel manner, prisoners in cells given a view of adjacent farmland had a lower sick-call rate than those confined in the same way but with a view of the prison courtyard. Similarly, business employees reported fewer feelings of stress and greater job satisfaction when their outside view was a natural environment.

In further support of human habitat selection, dental patients in sight of natural scenery registered lower blood pressure and reported reduced levels of anxiety. Psychiatric inpatients exposed to various forms of wall art responded most favorably to those depicting natural environments. During fifteen years of records of patient attacks on wall art, all were directed at abstract paintings, none expended on literalist representations of nature. (*Pace* abstract artists: this report is not a criticism; I know your purpose is often anything but quietude.)

While these several lines of evidence and others of similar import that have been adduced are only fragmentary in substance, they tell us that much of human nature was genetically encoded during the long stretches of time that our species lived in intimacy with the rest of the living world. People in most countries today have come to count that connection for little. They have pushed the rest of life to the margin, and rank its decline well down in the order of their personal concerns. But I believe that as the scientific study of human nature and living Nature grows, these two creative forces of the human self-image will coalesce. The central ethic will shift, and we will come full circle to cherish all of life—not just our own.

II
DECLINE AND REDEMPTION

BLINDED BY IGNORANCE

AND SELF-ABSORPTION, HUMANITY

IS DESTROYING THE CREATION. THERE IS

STILL TIME TO ASSUME THE STEWARDSHIP

OF THE NATURAL WORLD THAT WE OWE

TO FUTURE HUMAN GENERATIONS.

8

The Pauperization of Earth

PASTOR, YOU WILL KNOW ALREADY that according to fossil evidence and the best calculations by scientists, the last of the dinosaurs vanished suddenly from Earth 65 million years ago. Their extinction was part of an environmental Armageddon worthy of the book of Revelation. A giant meteorite, after burning through the atmosphere, crashed into the planet's surface. Its impact, in the vicinity of the present-day Yucatán peninsula of Mexico, threw mountainous tsunamis against the surrounding coasts, spewed dust high into the atmosphere. It rang Earth's crust like a bell, triggering volcanic eruptions around the world. The ejecta darkened the sky, altering the global climate. All these effects together rendered the land and sea uninhabitable for a majority of plant and animal species. Scientists have marked the event as the end of the Mesozoic Era, the Age of Reptiles, and the beginning of the Cenozoic Era, the Age of Mammals.

The end-of-Mesozoic extinction spasm had precedents. It was the fifth disturbance of this great magnitude in Earth's history during the 400 million years previous to that time. There were many smaller episodes in between, but the big five were the true history makers of life on Earth.

Schaus' swallowtail butterfly *Papilio aristodemus ponceanus*, a critically endangered race limited to one island in the Florida Keys. (From Susan M. Wells et al., eds., *The IUCN Invertebrate Red Data Book* [Gland, Switzerland: IUCN, 1984], p. 427.)

Now a sixth spasm has begun, this one a result of human activity. Although not ushered in by cosmic violence, it is potentially as hellish as the earlier cataclysms. According to estimates by a team of experts in 2004, climate change alone, if left unabated, could be the primary cause of extinction of a quarter of the species of plants and animals on the land by midcentury.

The list of species erased is already long. Since 1973, when Congress passed the Endangered Species Act to stem the hemorrhaging, over one hundred U.S. species have nevertheless vanished. Gone are the golden coqui, a Puerto Rican tree frog; the lotis blue butterfly of California; Bachman's warbler, a migratory species of the eastern United States; and all three of the land birds found uniquely on Guam, including the brilliantly colored cardinal honeyeater. The United States leads the world in the number of bird species lost during the past quarter century. The count is either five or seven, depending on whether two of the forms are classified as full species as opposed to mere geographical subspecies. Most of the losses occurred in Hawaii, America's notorious "extinction capital" and one of Earth's most biologically ravaged hot spots. Many other countries rival and probably far exceed the United States when all kinds of plants and animals are included. For example, 266 species of the exclusively freshwater fishes of peninsular Malaysia have been extinguished, as have 15 of the 18 unique fishes of Lake Lanao, in the Philippine Islands, and 50 species of the cichlid fishes of Africa's Lake Victoria.

The decline of Earth's biodiversity is an unintended consequence of multiple factors that have been enhanced by human activity. They can be summarized by the acronym HIPPO, with the order of letters corresponding to their rank in destructiveness.

H habitat loss, including that caused by human-induced climate change

I invasive species (harmful aliens, including predators, disease organisms, and dominant competitors that displace natives)

P pollution

P human overpopulation, a root cause of the other four factors

O overharvesting (hunting, fishing, gathering)

When a species declines toward extinction, not one but two or more factors are usually responsible. Thus overfishing in the

A recently vanished bird, the slender bush wren *(Xenictus longipes)* of New
Zealand. The last population was pushed to extinction by rats in the 1970s.
(From Tim Flannery and Peter Schouten, *A Gap in Nature: Discovering the
World's Extinct Animals* [New York: Atlantic Monthly Press, 2001], p. 169.)
Copyright © 2001 by Peter Schouten. Used by permission of
Grove/Atlantic, Inc.

sea with bottom drag nets (O) has simultaneously destroyed (H)
the sea floor habitat on which bottom species such as cod and
haddock depend. When an endangered bird or some other
species is restricted to a single small population by habitat
destruction (H), it becomes more susceptible to invasive predators and disease (I), pollution (P), and overharvesting (O). Much
of the science of conservation biology is devoted to the teasing
apart of these malign forces in order to weigh their importance,
then to nullify them.

A huge difference exists between temperate and tropical
regions in biodiversity loss. First, by far the greater part of biodi-

versity exists in the tropics: more than half the known species of Earth's plants and animals are confined to the rainforests alone. The pattern of loss also differs. During the past two millennia deforestation became severe first in the temperate countries. It spread from the Middle East and Mediterranean to Europe, thence to northern Asia, and on to North America. Finally, in the twentieth century, forest destruction swept through the tropics.

Now temperate forests have begun a limited regeneration, especially in Europe and North America, with an overall increase in cover of 1 percent during the 1990s. But tropical forests have continued to retreat, dropping 7 percent during the same decade. Between 1970 and 2000 the size of populations in temperate grasslands fell by 10 percent, as more arable land was developed. This was overwhelmingly surpassed during the same period by tropical grassland populations, which declined by a staggering 80 percent.

Freshwater ecosystems are pressed even more than forests and grasslands. Humans take up a quarter of the accessible water released to the atmosphere by evaporation and plant transpiration, and more than half the runoff from rivers and other natural channels. We are swiftly drawing down aquifers around the world, from America's Great Plains to China's Yellow River Basin and Saudi Arabia's irrigated desert. By 2025, some 40 percent of the world's population could be living in countries with chronic water scarcity. Of all Earth's water, marine included, fresh water makes up only 2.5 percent, and the greatest part of that is locked up in the planet's ice caps.

Not surprisingly, then, the highest rate of species endangerment per unit area occurs in freshwater ecosystems. There occurs a large fraction of Earth's biodiversity, including for example, 10,000 of the 25,000 known fish species. Many river systems approach the fate of those in China, where chiefly because of pollution 80 percent of the 50,000 kilometers of major channels can no longer support fish of any kind. Also, many lakes may go the way of Central Asia's Aral Sea. From 1960 to 2000 its area shrank by one-half, because of the blockage of the Amu Darya and Syr Darya rivers. Its salinity has increased almost fivefold, and its fisheries have collapsed. Among the collateral Aral Sea catastrophes, 159 bird and 38 mammal species have disappeared from the two river deltas.

In shallow tropical water coral reefs, the biologically rich "rainforests of the sea," are retreating worldwide, variously bleached by climatic warming, polluted, and dynamited to harvest fish, split by artificial channels, and excavated for building materials. Those around Jamaica and some other Caribbean islands have largely disappeared. Even the Great Barrier Reef of Australia, the largest and best protected in the world, declined 50 percent in cover between 1960 and 2000. Overall, about 15 percent of the world's coral reefs are gone or judged to be damaged beyond repair, and another third could be lost during the next thirty years if the present downward trends continue.

Not even the high seas are safe from the human onslaught. Fish that exist at high trophic levels and hence are large and commercially favored, such as cod and tuna, declined precipitously through overharvesting between 1950 and 2000.

Although ecosystem destruction can be measured with some confidence by remote sensing and ground surveys, it remains notoriously difficult to estimate rates of species extinction. By extinction is meant the confirmed disappearance of every last individual, everywhere. Some animals, such as the larger birds and mammals, in particular among them the slow and tasty, are more susceptible to extinction than most other organisms. Gone, for example, are the elephant bird of Madagascar, all the ostrich-like moas of New Zealand, and the vast majority of North American mammal species over ten kilograms in weight. The same is true of fishes originally limited to one or two freshwater streams. Most kinds of insects and other small organisms are still so difficult to identify and monitor as to prevent precise censuses. Still, biologists using several indirect methods of analysis generally agree that at least on the land and in freshwater ecosystems, ongoing extinctions are very roughly 100X higher than before the arrival of modern *Homo sapiens* about 150,000 years ago. The 100X figure is an order-of-magnitude, or nearest powers-of-ten, estimate. The extinction rate, in other words, is likely more than 50 times and less than 500 times the prehuman baseline. The rate is all but certain to rise to order of magnitude 1,000 or even 10,000, as species now rated endangered die off, and the last remnants of some ecosystems are destroyed, sweeping away the species limited to them.

Conservation biologists have lately paid special attention to the plight of the world's 5,743 known species of amphibians, comprising frogs, toads, and salamanders along with caecilians, a small group of tropical species with serpentine bodies. A marked decline during the past three decades is thought by many experts to foreshadow a similar drop in the rest of global plant and animal biodiversity.

The first signs of the amphibian crisis were detected more or less simultaneously in scattered parts of the world during the 1980s. In the next decade species extinction in frogs and toads in particular was recognized as a major environmental problem and given a name, the Declining Amphibian Phenomenon. In 2004 an international team of amphibian experts reported the results of a study covering the preceding several years: 32.5 percent of amphibian species worldwide were classified as threatened with extinction, compared with 12 percent of reptiles, 23 percent of birds, and 23 percent of mammals. Many had been designated "critically endangered" in the Red List of the International Union for the Conservation of Nature. Thirty-four amphibian species have been confirmed extinct (9 since 1980), and 113 other species classified as "possibly extinct" since 1980. Of the last, no specimens could be found, yet the species will not be considered formally extinct until searches conducted over extended periods of time prove fruitless.

This ongoing biological catastrophe, and it can scarcely be called anything less than that, is dramatically illustrated by the condition of the amphibians of Haiti. This small Caribbean nation has stripped away all but 1 percent of its forests and thoroughly polluted its streams and rivers. In a land once celebrated for its lush tropical landscape and rich fauna and flora, the very existence of 47 of the 51 known Haitian amphibian species is now threatened. Of these, 31, or two-thirds of the total, are classified as critically endangered, subject to complete extinction in the near future. Ten are thought merely "endangered" and 5 "vulnerable."

Loss of habitat and pollution are manifestly the primary reasons for the decline in the Haitian amphibian fauna. Elsewhere these lethal forces unleashed by man are at work singly or, more likely, in combination with others. All are the unintended result of human activity. Habitat loss is mostly responsible for decline and extinction of amphibians in the western United States,

Spain, West Africa, and Indonesia. Habitat loss exacerbated by the restricting effects of climatic change causes the most damage in montane Central America and the Atlantic forest of Brazil. The spread of a fatal chytrid fungus has been a key factor in Central America and tropical northeastern Australia, while overharvesting of frogs is the primary agent in mainland southeastern Asia.

Kermit the Frog, to summarize the situation in a phrase, is sick. And to varying degrees so is much of the rest of the living world. Might *Homo sapiens* follow? Maybe, maybe not. But with certainty we are the giant meteorite of our time, having begun the sixth mass extinction of Phanerozoic history. We are creating a less stable and interesting place for our descendants to inherit. They will understand and love life more than we, and they will not be inclined to honor our memory.

9

Denial and Its Risks

DEAR PASTOR, what I fear most is the pervasive combination of religious and secular ideology of a kind that sees little or no harm in the destruction of the Creation. The following speech might be given by the visionary who ranks biodiversity of little account and sees humanity ascending profitably away from Nature and not to it. He says to those who wish to save wild Nature:

Brothers and sisters, do not weep, for what will soon pass from Earth. Life is change, and extinction is sometimes a good thing. Instead, celebrate humanity as a new order of life, and the "plundered" planet as the new biosphere. Let any species that blocks progress slip away. Before the coming of mankind, there was always a turnover of ecosystems and species. Even if the world is biologically impoverished in furthering the interests of humanity, our species is in no danger. When one resource is exhausted, our scientific and technological genius will find another.

Look to space, my good people. Look to the heavens! Do not think vanished faunas and floras a bitter heritage for future generations. We can keep some nature parks the way we preserve historic old buildings, to remind us of the past. Perhaps we will even create new ecosystems with

advanced bioengineering and stock them with species of our own mak-
ing. Who knows what wonderful creatures will be fashioned? They
would be works of art, ever more aesthetically pleasing and useful in
many ways. A prosthetic and superior environment will replace the old
and primitive.

It is within the power of future technology, perhaps in accord with
divine providence, for people to flourish as never before in a completely
humanized environment, a paradise of our own making. Such is the
foreordained trajectory of an advanced intelligent species. I tell you, it is
our destiny! In coming generations medicines will be synthesized from
chemicals off the shelf, food grown from a few dozen genetically enhanced
crop species, and the atmosphere and climate controlled by computer-
guided sustainable energy sources. This old Earth will go on spinning
through space as it has for billions of years (or, if you prefer, six thousand
years). The planet will become a literal, not just a metaphorical, space-
ship. Our finest minds will be up there on the bridge of voyaging Earth,
reading monitor displays, touching buttons, keeping us safe.

Such is the philosophy of exemptionalism, which supposes
that the special status on Earth of humanity lifts us above the
laws of Nature. Exemptionalism takes one or the other of two
forms. The first, just expressed, is secular: don't change course
now, human genius will provide. The second is religious: don't
change course now, we are in the hands of God, or the gods,
Earth's karma, whatever.

A cheerful faith in human destiny dismisses the rest of life
through successive denials. The first says, Why worry? Extinction is natural. Life forms have been dying out over billions of
years of history without any clear harm to the biosphere. New
species are constantly being born to replace them.

All this is true as far as it goes, but with a terrible twist. Except for giant meteorite strikes or other catastrophes every 100 million years or so, Earth has never experienced anything like the contemporary human juggernaut. With the global species extinction rate now exceeding the global species birthrate at least a hundredfold, and soon to increase to ten times that much, and with the birthrate falling through the loss of sites where evolution can occur, the number of species is plummeting. The original level of biodiversity is not likely to be regained in any period of time that has meaning for the human mind.

The second stage of denial takes form in a question, Why do we need so many species anyway? Why care, especially when the vast majority are bugs, weeds, and fungi? An exemptionalist religious scholar might add that an immense array of creatures discovered by science, including encytraeids, nematodes, rotifers, gnathostomulids, oribatids, archaea, and many others, are not even mentioned in Holy Scripture. It is easy to look past these creepy-crawlers, forgetting that only a century ago, before the rise of the modern conservation movement, native birds and mammals were eliminated with equal disregard. In just four decades, the population of passenger pigeons plunged from hundreds of millions to zero. The beautiful crimson-and-green Carolina parakeet changed from an abundant orchard pest to a receding memory. The bison of North America, and its European cousin the wisent, came within a few hundred rifle shots of extinction. Only now are they recovering, and then only in part. People today understand what was lost or almost lost in these cases by the unintended consequences of human greed. In time they will come similarly to value other creatures that still fall below their notice.

People will more widely share the knowledge acquired by biologists that these often obscure life forms run Earth completely free for us. Each is a masterpiece of evolution, exquisitely well adapted to the niches of the natural environment in which it occurs. The surviving species around us are thousands to millions of years old. Their genes, having been tested each generation in the crucible of natural selection, are codes written by countless episodes of birth and death. Their careless erasure is a tragedy that will haunt human memory forever.

Even if that much is granted, the third stage of denial predictably emerges: Why rush to save all of biodiversity *now*? We have more important things to do. Priority is owed economic growth, jobs, military defense, democratic expansion, alleviation of poverty, medicine. Why not collect or gather live specimens of every species, and breed them in zoos, aquaria, and botanical gardens, for later return to the wild? Yes, this rescue

Old Blue, last surviving female of the Chatham Islands black robin and progenitor of the still-surviving species. (Photograph by Don Mertan. From David Butler and Don Merton, *The Black Robin* [New York: Oxford University Press, 1992], p. 149.)

operation is available as a last resort, and has in fact saved a few plants and animals that were on the brink of extinction. The successes deserve celebration and accolades, so let me pause to tell you about them. The most spectacular example is the black robin of the Chatham Islands, an archipelago east of New Zealand. By 1980 rats and feral cats introduced by settlers had reduced the once abundant robins to just one breeding pair. Kept in captivity, "Old Blue" and "Old Yellow" mated and reared young, and their descendants have now been used to repopulate some of the original habitat on two of the islands. That was the closest call in conservation history.

A second Lazarus project brought back the Mauritian kestrel, a small tawny falcon limited to the same Indian Ocean island that once harbored the dodo, the world's icon of extinction. By 1974, pesticide contamination of the environment had reduced the wild peregrine population to four individuals. Like the last Chatham Islands black robins, captive birds multiplied, and today their descendants swoop through the remnant woodlands that line the Mauritian ravines. Other Lazarus species are the California condor, the bird with the widest wingspan of all American species, which after captive breeding has been returned to the wild in the Grand Canyon; the beautiful Pere David's deer, limited to zoos and parks after it had been hunted to extinction from the marshlands and forests of northeastern China (and soon to be returned there); the Laysan duck of the Hawaiian Islands, up from seven survivors and currently hold-ing well at five hundred adults; and the whooping crane, a majestic presence of the North American heartland, down to fourteen adults and all but given up as lost in 1937, but up now to a population exceeding two hundred.

An ivory-billed woodpecker. (From James C. Greenway Jr., *Extinct and Vanishing Birds* [New York: American Committee for International Wildlife Protection, 1958], p. 358.)

A new and already world famous Lazarus candidate is the ivory-billed woodpecker, a dramatically large, conspicuous bird of the southern United States. The Lord God bird, as it was sometimes locally called (some people would say when they first saw one, "Lord God, what is that?"), was thought to have gone extinct in 1944, when the last known individual was spotted in the newly cut-over Singer tract of Louisiana. In the years to follow, birders searched for ivorybills in the sparse remnants of its favored habitat, the old-growth bottomland forests. Occasional rumors of sighting were heard—they were favorite items of gossip among naturalists—but none was substantiated. As hope faded, the ivorybill became the Holy Grail of ornithology, a legendary being, pursued only by the obsessed. Then, in the spring of 2005, came some electrifying news: a male had been spotted the year before in the floodplain forest of the Cache River Wildlife Refuge, of eastern Arkansas, and then quietly verified by experts by eight more sightings. Its red peaked crown and white primaries are evident in photographs and a videotape. The number surviving must be very small, given the five to fifteen square miles of old-growth forest needed to support a single pair. The Cache River reserve might support twenty to sixty pairs under optimal conditions. Yet, it is equally possible that all the sightings to date in 2005 are of only a single individual.

The successes of last-ditch efforts and the occasional rediscovery of supposedly lost species should not beguile us into thinking that in time we will see a great deal of lost biodiversity return in the diminished space we have left to Nature, such as the Cache River bottomlands. To make this point clear, it is necessary only to list the bird species native to the United States that have disappeared during the past quarter century, with the date of their last sighting. Most are island species, and two, the mallard and sparrow, may be of less than species rank: olomao (1980), Mariana mallard (1981), Guam flycatcher (1983), kamao (1985), Oahu alauahio (1985), Kauai'oo (1987), dusky seaside sparrow (1987), ou (1989), poouli (2005). Because most of these species were limited from the start to small geographical ranges, in contrast to the ivory-billed woodpecker, there is much less chance they still survive.

Successful recoveries of critically endangered species will of necessity continue to be rare exceptions. So we come back to the Lazarus dream. The sobering truth is that all the zoos in the world can sustain breeding populations of a maximum of only two thousand mammal species, out of about five thousand known to exist. A similar limitation exists for birds. Botanical gardens and arboreta are more capacious, but would be overwhelmed by the tens of thousands of plant species needing protection. The same is true of fishes that might be saved in aquaria. A lot of good can be accomplished, but at considerable expense per species, and it can only make a dent in the problem.

And how are we even to think of such an emergency measure for the millions of species of insects and other invertebrates, most still unknown to science—and still more, the tens of millions of microorganisms?

There is no solution available, I assure you, to save Earth's biodiversity other than the preservation of natural environments in reserves large enough to maintain wild populations sustainably. Only Nature can serve as the planetary ark.

So here, Pastor, is a homily of my own I offer to counter that of the exemptionalist:

Save the Creation, save all of it! No lesser goal is defensible. However biodiversity arose, it was not put on this planet to be erased by any one species. This is not the time, nor will there ever be a time, when circumstance justifies destroying Earth's natural heritage. Proud though we are of our special status, and justifiably so, let us keep our world-changing capabilities in perspective. All that human beings can imagine, all the fantasies we can conjure, all our games, simulations, epics, myths, and histories, and, yes, all our science dwindle to little beside the full productions of the biosphere. We have not even discovered more than a small fraction of Earth's life forms. We understand fully no one species among the millions that have survived our onslaught.

It is true that nonhuman life preceded us on this planet. Whether by a literal day, according to Genesis, or by more than 3.5 billion years, as the scientific evidence shows, it is still true that we are a latecomer. The biosphere into which humanity was born had its Nature-born crises, but it was overall a beautifully balanced and functioning system. It would have continued to be so in the absence of Homo sapiens. *Even today a diminished wild Nature provides us ecosystem services, such as water management, pollution control, and soil enrichment, equal in economic value to all that humanity artifactually generates.*

Think of it. With the smaller population that can be reached within a century, and a higher and sustainable per capita consumption spread more evenly around the world, this planet can be paradise. But only if we also take the rest of life with us.

1 0

End Game

THE HUMAN HAMMER having fallen, the sixth mass extinction has begun. This spasm of permanent loss is expected, if it is not abated, to reach the end-of-Mesozoic level by the end of the century. We will then enter what poets and scientists alike may choose to call the Eremozoic Era—the Age of Loneliness. We will have done it all on our own, and conscious of what was happening. God's will is not to blame.

The first five spasms took ten million years on average to repair by natural evolution. A new ten-million-year slump is unacceptable. Humanity must make a decision, and make it right now: conserve Earth's natural heritage, or let future generations adjust to a biologically impoverished world. There is no way to weasel out of this choice. I've explained why the zoo-and-garden option won't work. Knowing that, some quixotic writers have toyed with the idea of last-ditch measures. They say, Let's conserve the millions of surviving species and races by deep-freezing fertilized eggs or tissue samples for later resurrection. Or, let's record the genetic codes of all the species and try to re-create organisms from them later. Either solution would be high-risk, enormously expensive, and, in the end, futile. Even if

Earth's threatened biodiversity in all its immensity could be reanimated and bred into populations awaiting return to what might in the twenty-second century pass for the "wild," the reconstruction thereby of independently viable populations is beyond reach. Biologists haven't the slightest idea of how to build a complex autonomous ecosystem from scratch. By the time they find out, they may discover that conditions on the humanized planet make such a reconstruction impossible.

Passing beyond these options leaves a final one for the exemptionalists to pose: go ahead and pauperize the biosphere, in the hope that scientists may someday be able to create artificial organisms and species and put them together in synthetic ecosystems. Let future generations refill the niches of vanished Nature with tigeroids programmed not to attack humans, synthetic tigers burning artificial bright in forestoids amid insectoids that neither sting nor bite. There are words appropriate for artifactual biodiversity, even where it exists only in fantasy: desecration, corruption, abomination.

All of the aforementioned default solutions have been suggested, I am sorry to say, by one writer or another. The dreams are fatuous. This is the time not for science fiction but for common sense and the following prescription: ecosystems and species can be saved only by understanding the unique value of each species in turn, and by persuading the people who have dominion over them to serve as their stewards.

Humanity is in a bottleneck of overpopulation and wasteful consumption that can open out by the end of the century, when the global population is expected to peak at around nine billion, 50 percent more than what it was in 2000, then commence to recede. During the remainder of the bottleneck period, per capita consumption will also rise, increasing pressure on the environment. But it too can be brought under control, in large part by already existing technology that raises production while recycling materials and converts to alternative energy sources. This shift seems inevitable anyway because of a corporate-level Darwinism: those corporations and nations committed to further improvement and application of the technology will be the economic leaders of the future.

If we wish, a greater part of the ecosystems and species that still survive can be brought through the bottleneck. The methods to save them exist. They are being applied at local and national levels around the world, albeit still sporadically. The ongoing effort is still far from enough to save the bulk of species that have reached the critically endangered level. But it is a beginning, and one widely understood and approved. The commitment of independent nations to take action is growing rapidly. By 2002, 188 had signed the Convention on Biodiversity initiated ten years earlier at the Rio Summit (the United States, ideologically isolationist in all matters except commerce, tourism, and democratic expansion, was and remains a nonsignatory; the others holding back, at this writing in 2006, are Andorra, Brunei, Iraq, Somalia, East Timor, and the Vatican). Meeting in Johannesburg, the signatories pledged cooperative action to reduce the rate of biodiversity loss significantly by 2010. At the same time, 130 of the 191 UN members, again not including the United States, have amended their constitutions to protect their national environment, in most cases directly or implicitly including biodiversity.

A race is now on that will decide the fate of the greater part of Earth's biodiversity. The choice is simple: save biodiversity during the next half century or lose a quarter or more of the species. Realization that this Armageddon can be quickly won, or lost, is

Thirty-four of the most critical biodiversity hot spots on land: geographical areas with large numbers of endangered species. (© 2006, Conservation International. All rights reserved.)

based on knowledge of the geography of life, a key principle of which is that species do not occur evenly over the land and sea, but in concentrations called hot spots. You are far more likely, for example, to find an endangered species in an upland Florida scrub savanna than in a Wisconsin woodland, or in a North Carolina mountain stream than in a New Hampshire river. The hottest of the hot spots, those in most critical need of immediate attention, are scattered around the world, sometimes in surprising locations. Those on the land identified by Conservation International in 2006 include the following:

- California's coastal and foothill sage
- The tropical forests of southern Mexico and Central America
- The forests and dryland habitats of the Caribbean islands, especially Cuba and Hispaniola
- The tropical lowland and midlevel forests of the Andes
- The cerrado (savanna) of Brazil
- Brazil's Atlantic forest
- The forests and dryland habitats of the Mediterranean Basin
- The forests of the Caucasus Mountains
- The Guinean forests of West Africa
- Multiple habitats of the Cape region, southern Africa
- Multiple habitats in the Horn of Africa
- Multiple habitats, but especially the forests, of Madagascar
- The rainforests of India's Western Ghats
- The rainforests of Sri Lanka
- The forests of the Himalayas
- The forests of Southwest China
- Most of the forests of Indonesia
- The rainforests of the Philippines
- The heathland of Southwest Australia
- The forests of New Caledonia
- The forests of Hawaii and many other eastern and central Pacific archipelagoes

Thirty-four of the hottest spots, or more precisely the intact biologically rich habitats within them, cover a mere 2.3 percent of Earth's land surface, yet they are the exclusive homes of 42 percent of the planet's vertebrate species (mammals, birds, reptiles, and amphibians) and 50 percent of its flowering plants.

The hot spots are not merely points of concentration of bio-diversity. They are by virtue of their limited area the location of many of the planet's most vulnerable species. A large majority of the species classified in the Red List of the International Union for the Conservation of Nature as "endangered" or "crit-ically endangered" live within the thirty-four hottest spots, including 72 percent of Earth's mammals, 86 percent of the birds, and 92 percent of the amphibians.

Species are the preferred unit of measurement of biodiversity because they are by and large natural units in evolution. They can be more precisely delimited than ecosystems, and they are easier to identify than the complex ensembles of genes that dis-tinguish them from other species.

Species have one disadvantage as units in the measurement of biodiversity: they often occur in clusters that have evolved recently, in extreme cases within a few thousand years. Because of their youth, species in these "sibling" clusters tend to differ relatively little from one another in genetic composition. Is there a way of measuring biodiversity by counting whole clusters rather than the species composing them? There is a way, one that dates back to the beginnings of formal taxonomic nomen-clature in the mid-eighteenth century. In the hierarchical system used, a cluster of species that are similar in their distinguishing traits, hence likely to be close genetic relatives, are classified as a genus. These genera (plural of genus) are thus older, more diver-gent assemblages and can be used instead of relatively "cheap" species to cut farther back in time to assess biodiversity. When that is done, do the hot spots change? The answer is yes but not much; they remain mostly the same as those based on species alone. However, their rank order shifts, and in the following manner. The hottest of the hot spots on Earth, far out front with 478 genera of plants and vertebrates all its own, is Madagascar, the ancient great island off the east coast of Africa. Following Madagascar (with the number of exclusively owned genera added in parentheses) are the Caribbean islands (269), the Atlantic Forest of Brazil (210), the Sunda archipelago of Indonesia (199), the mountains of East Africa (178), the South African Cape (162), and southern Mexico plus Central America (138).

Most of the early hot-spot studies were limited to land environments. Since 2000, similar modes of analysis have been applied to marine environments. Three of the four major zones, namely estuaries, coral reefs and other shallow-water habitats, and the floor of the deep sea, are fragmented into small and often threatened places in a manner similar to the hot spots on the land. The fourth marine zone, the high seas, also varies in biological richness from one part of the globe to another, but its patterns are hard to pin down, because of the readiness with which so many oceanic fish and other open-water organisms travel long distances.

To summarize to this point, the results of global biodiversity studies are now sufficient for a successful application to conservation practice. Biologists have put a measure on the size of the problem. They can project many of the consequences that will follow if the trends are not abated. They know how to fix the problem, at least most of it.

With all that in mind, let us move to the bottom line: How much will it cost to fix the problem? It may be feared that saving biodiversity will be so expensive as to endanger the economy, that is, the market economy. This assumption is a mistake. The cost of saving most of Earth's fauna and flora would be relatively trivial for the market economy and, of course, immensely profitable for the natural economy. In 2000 Conservation International sponsored a conference of biologists and economists, entitled "Defying Nature's End," to address this matter. They reviewed the many methods available at that time to secure wildland reserves while simultaneously improving local economies, then estimated the cost. They concluded that in order to put a protective umbrella over the twenty-five hottest spots on the land then recognized (nine more have since been added to total the aforementioned thirty-four), plus core areas within the remaining tropical forest wildernesses, those of the Amazon and Congolian basins and New Guinea, would require one payment of about $30 billion. The benefit, if the allotment is joined with wise investment strategy and foreign policy, would be substantial protection for 70 percent of Earth's land-dwelling fauna and flora. It would at least give time to devise new methods and new policies for the long term. This *single* outlay (one payment only), or its equivalent spread over a few years, is approximately one part in a thousand of the *annual* gross world product, that is, gross domestic products of all countries combined. By coincidence the latter amount, roughly $30 trillion, also happens to be the estimated rate of the ecosystems services given free by Earth's remaining natural environment.

A parallel study, made in 2004 by a second team, estimated the cost of protecting marine areas, the threatened Second Edens of our planet. They recognized that the oceans can no longer be treated as limitless or invulnerable. Coral reefs are retreating by physical destruction and the ill effects of climate warming. All of the major open-ocean fisheries are operating below sustainable levels. And much of the shallow ocean floor around the world has been ruined by bottom trawling. Existing marine reserves within the 370-kilometer exclusive economic zones of the coastal nations cover only 0.5 percent of the ocean surface, and except for restrictions on whaling, there is no protection at all for life of the high seas. If reserves were set up over the whole of the coastal zones and open seas and expanded sufficiently in area, the result would be security for countless threatened species. In time the reserves would also raise the sustainable yield of the fisheries, by serving as sources for wide-ranging marine organisms. To regulate a reserve network covering 20–30 percent of the ocean surface would cost between $5 billion and $19 billion annually. That outlay could be met by eliminating the current perverse subsidies given to the fishing industry, which fall between $15 and $30 billion annually—and are responsible in the first place for the overharvesting and falling yield of preferred species.

Life on this planet can stand no more plundering. Quite apart from obedience to the universal moral imperative of saving the Creation, based upon religion and science alike, conserving biodiversity is the best economic deal humanity has ever had placed before it since the invention of agriculture. The time to act, my respected friend, is now. The science is sound, and improving. Those living today will either win the race against extinction or lose it, the latter for all time. They will earn either everlasting honor or everlasting contempt.

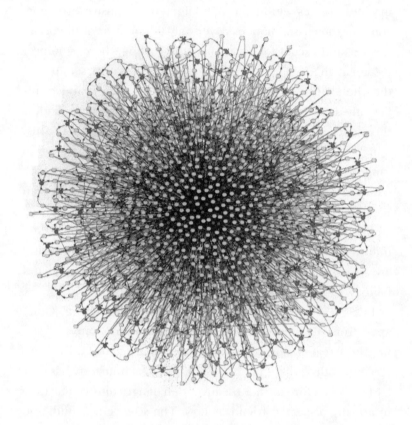

III
WHAT SCIENCE
HAS LEARNED

ARGUMENTS FOR SAVING

THE REST OF LIFE ARE DRAWN

FROM BOTH RELIGION AND SCIENCE.

THE RELEVANT PRINCIPLES OF BIOLOGY,

THE KEY SCIENCE IN THIS DISCOURSE,

ARE EXPLAINED HERE.

11

Biology Is the Study of Nature

IN MY OPINION, PASTOR, the ascent to Nature and the restoration of Eden do not need more spiritual energy. Of that people have a superabundance. Rather, spiritual energy must be more broadly applied, and more exactly guided by an understanding of the human condition. Humanity's self-image has risen far during the past three hundred centuries. First lifted by religion and the creative arts, it can rise still higher on the wings of science.

To support that claim, I will offer now an account of the concept and practice of science and in particular biology, the discipline most immediately relevant to human concerns.

I hasten to add I do not mean *scientists*. Most researchers, including Nobel laureates, are narrow journeymen, with no more interest in the human condition than the usual run of laymen. Scientists are to science what masons are to cathedrals. Catch any one of them outside the workplace, and you will likely find someone leading an ordinary life preoccupied with quotidian tasks and pedestrian thought. Scientists seldom make leaps of the imagination. Most, in fact, never have a truly original idea. Instead, they snuffle their way through masses of data and hypotheses (the latter are educated guesses to be tested), some-times excited but most of the time tranquil and easily distracted by corridor gossip and other entertainments. They have to be that way. The successful scientist thinks like a poet, and then only in rare moments of inspiration, if ever, and works like a bookkeeper the rest of the time. It is very hard to have an original thought. So for most of his career, the scientist is satisfied to enter the figures and balance the books.

Scientists are also like prospectors. Original discoveries are the gold and silver of their trade. If important, they can buy collegial prestige, and with it wider fame, royalties, and academic tenure. Scientists by and large are too modest to be prophets, too easily bored to be philosophers, and too trusting to be politicians. Lacking in street smarts, they are also easily fooled by confidence artists and sleight-of-hand tricksters. Never ask a scientist to test the claims of paranormal phenomena. Ask a professional magician.

The power of science comes not from scientists but from its method. The power, and the beauty too, of the scientific method is its simplicity. It can be understood by anyone, and practiced with a modest amount of training. Its stature arises from its cumulative nature. It is the product of hundreds of thousands of specialists united by the one binding commonality of the scientific method. Few scientists know more than a small fraction of available scientific knowledge, even within their own disciplines. But no matter: their fellow scientists are continuously testing and adding to the other parts, and the entire body of scientific knowledge is easily available. The invention of this remarkable engine of testable learning was the one advance in recorded human history that can be called a true quantum leap. But it attained its preeminence relatively late in the geological life span of humanity, and only after the human intellect had traveled a long, tortuous path dominated by tribalism and animated by religion.

Let's try to establish a rough chronology. Millions of years ago there was only animal instinct. Then, probably at the man-ape level, the rudiments of materials culture were added. With still higher intelligence there followed a sense of the supernatural, whereupon demons, ancestral ghosts, and divine spirits peopled the human mind. Without science there had to be religion, in order to explain man's place in the universe. Born of dreams, its images were enshrined in the culture by shamans and priests. The gods made man. Those that lived in surrounding Nature gave way to gods on sacred mountains, in distant places, and in the heavens. Somewhere and somehow back in time, these divine humanoids had created the world, and now they governed man. Humans, in their evolving self-image, rose above Nature to follow the gods as children and servants. Tribes led unwaveringly by their personal gods were united and strong. They defeated competing tribes and their false gods. They also subdued Nature, erasing most of it in the process. Their destiny, they believed, was not of this world. They thought of themselves as immortal, no less than demigods.

Along the way, commencing in Europe in the seventeenth century, a radical alternative self-image emerged. Art and philosophy began to disentangle themselves from the gods, and science learned to operate with full independence. Step by step, often opposed by the followers of Holy Scripture, science constructed an alternative worldview based on a testable and self-reliant human image. Doubling in growth every fifteen years during most of the past three and a half centuries, it has looked into the heart of living Nature, finding there a previously unimagined vast and autonomous creative force. This image has subsumed religious rivalries and reduced them to intertribal conflict. Science has become the most democratic of all human endeavors. It is neither religion nor ideology. It makes no claims beyond what can be sensed in the real world. It generates knowledge in the most productive and unifying manner contrived in history, and it serves humanity without obeisance to any particular tribal deity.

Biology now leads in reconstructing the human self-image. It has become the paramount science, exceeding other disciplines, including physics and chemistry, in the creative tumult of its discoveries and disputations. It is the key to human health and to management of the living environment. It has become foremost in relevance to the central questions of philosophy, aiming to explain the nature of mind and reality and the meaning of life. Not least, biology is the logical bridge between the three great branches of learning: the natural sciences, the social sciences, and the humanities.

Individual scientists, whose professional ethic is founded in objectivity, are usually careful not to overstate their ambitions in public. Nevertheless, from the essays and lectures of its more audacious leaders, it is possible to glean a picture of the great goals of present-day biology. They are, I believe, as follows:

- Create life: complete the mapping of a species of simple bacterium at the molecular level, simulate its processes by computers, then construct individual bacteria from the constituent molecules, or at least show how such construction can be accomplished.

- Using this approach and combining it with knowledge of the chemistry of early Earth, reconstruct the steps that led to the origin of life.
- Continuing to advance the same molecular reduction and synthesis to human cells, use the information with increasing effectiveness to cure disease and repair injuries.
- Explain the mind with models of chemical and electrical transmission and the molecular basis of nerve-cell growth and network formation; then simulate the mind with the combination of artificial intelligence and artificial emotion.
- Complete the mapping of Earth's fauna and flora to the species level, including microorganisms, and expand exploration of diversity at the gene level for each of the species.
- Use the exponentially growing information about diversity within the biosphere to advance medicine, agriculture, and public health.
- Create a Tree of Life for all species and for major gene ensembles within them, thereby tracing the pathways of past evolutionary histories. Meanwhile, combining this information with paleontology and environmental history, establish definitive principles concerning the origin of biodiversity.
- Decipher how stable natural communities are assembled and regulated at the species level; use this information to protect and stabilize Earth's biodiversity.
- Bridge, if not outright unify, the natural sciences, social sciences, and humanities by exploration of the biological foundation of mind and human nature. In the process, unveil the coevolution of genes and culture.

Measured against this imagined scenario of its ultimate maturity, present-day biology is a still primitive science in comparison with physics and chemistry. How, then, might it grow to match its visions?

Consider first how the science is constructed. Biology is a science of three dimensions. The first dimension is the study of individual species (thus, one kind of bacterium or one kind of fruit fly) across all the hierarchical levels of biological organization the species have attained—from molecules to the cells the molecules compose and energize, to the tissues and organs constructed from the cells, to the organisms made up of tissues and organs, and on to societies and populations of organisms, and finally to the interactions of the species to form ecosystems.

Species are genetically distinct populations that in many, but far from all, kinds of organisms are separated by their inability to crossbreed in natural environments. All the species that live in a particular pond or forest, for example, are the living community. In combination with the nonliving soil, air, and water, these species compose the ecosystem.

The first dimension of biology, to repeat, is the scan of individual species all the way from their molecular composition to their place in ecosystems. The second dimension is the mapping of biological diversity, "biodiversity" for short, of all the species in a selected part of the world, whether a local habitat, or region, or all the planet, together with the ecosystems the species form and the genes that prescribe the diagnostic traits of the species. The third dimension of biology is the history of each of these species, ecosystems, and genes. Ecologists track species through seasons and generations in order to understand how their populations wax and wane. By vastly expanding the scale of inquiry, systematists and geneticists reconstruct history through enough generations to witness changes in the genes and, at a higher level, the splitting of the species into daughter species.

Try now to envision simultaneously the reach of the three dimensions of biology. You cannot, I cannot, no one can—at least not yet. Uncounted millions of species exist. The great majority are still unknown to science. Examined in one slice of time, each species is a unique creation: its genetic code has been fashioned through an almost unimaginably complex trajectory of mutations and natural selection that led to its currently defining traits.

Each species is a world in itself. It is a unique part of Nature. In the instant of time the species comes to your attention, it is spread before you as an ensemble of its member organisms, distributed in certain patterns over the landscape. In your imagination let the clock run, then speed it up, faster and faster. The constituent organisms disperse and die as new ones are born; they too disperse and die; and so on until the whole population of organisms declines to extinction. The dynamics of the population are controlled by changes in the environment, by heavy rains or a drought, encroachment or retreat of pathogens and predators, abundance or shortage of food, and more. These factors and the way they influence one another cause the species to expand or contract, to penetrate new habitats, or to tumble to extinction.

Finally, in your mind try to overlay millions of such species evolving forward in time and then far back into the history of each of them, at all levels of organization from gene to ecosystem. There in a nutshell is the transcendent and only dimly foreseeable complexity of future biology. There is to be found a new theater of spiritual energy.

12

The Fundamental Laws of Biology

LET ME NEXT PRESENT the subject in a different way. The most efficient way to grasp the importance of biology for the human condition is to address the science from the top down—first its most general laws, then in diminishing blocks the particularities governed by them.

A law in biology is the abstract description of a process that evidence suggests is universal in living systems and possesses a logical inexorability for those systems. Scientists have settled upon what can arguably be called the two fundamental laws of biology. The first is that all the known properties of life are obedient to the laws of physics and chemistry. This is not meant to imply that all the properties can be explained directly by physics and chemistry. It means only that when the complex machinery of life is cleaved into its elements and processes, these parts and what is known of the interaction among them conform to what is known of physics and chemistry.

The division of a cell seen through a visible-light microscope does not surrender directly to a physicochemical explanation. We cannot see the physical and chemical processes directly. But the molecules that compose the cell and the choreography of their duplication do yield. The properties of the cell as a whole are called "emergent," meaning they arise from the interactions of the molecules. But because of the large number and complexity of the processes at that level, the movements cannot be readily deduced from principles of physics and chemistry. Hence until the interactions are worked out in detail, a step most likely to be achieved only with the aid of mathematical models and supercomputer simulation, cell division must be described in part at the cellular level, with a language different from that of physics and chemistry.

An emergent property, then, can be defined as one so complex and poorly understood that it must be described by an imagery and a vocabulary different from those used for the processes that create it. Most of biology consists of emergent properties, and for the time being therefore can be only loosely connected in causal explanation to physics and chemistry.

The crucial link between biology and the physical sciences is the structure of DNA, the molecule that encodes heredity. In 1953 James D. Watson and Francis H. C. Crick provided the chemical structure of this "key of life." Perhaps I draw too fine a point on the matter, but the following three sentences from the Watson-Crick report may fairly be said to have given birth to molecular biology, and thus decisively vindicated the first law of biology:

> We wish to put forward a radically different structure for the salt of deoxyribose nucleic acid. This structure has two helical chains each coiled around the same axis. . . . It has not escaped our notice that the specific pairing we have postulated immediately suggests a possible copying mechanism for the genetic material.

Today, molecular and cellular biology, the disciplines that enjoy the greatest support and activity from society, continue to address the two lowest levels of biological organization, molecule and cell. They focus on a few species selected for their special traits. For example, the colon bacterium *Escherichia coli* (usually, *E. coli*) is important for its genetic simplicity and ultrashort generation time, the roundworm *Caenorhabditis elegans* (usually, *C. elegans*) for the simplicity of its nervous system and behavior, and of course human beings, for which virtually every scrap of information has both fundamental and practical value.

Molecular and cellular biology are in the natural history period of their development. This perhaps surprising characterization can be clarified with a metaphor. The cell is a system consisting of a very large number of interacting elements and processes. In a fundamental sense it equates to an ecosystem such as a pond or forest. The molecules that make up the cell are the equivalent of the plants, animals, and microbes that compose the living part of an ecosystem. The two levels, cell and ecosystems, have been about equally well explored to the present time. The molecular and cellular biologists are discovering vast arrays of proteins and other molecules.

These researchers are the Humboldts, the Darwins, and other explorer-naturalists of a new age. Working in laboratories, mercifully free of mosquito bites and blistered feet, they press into the unmapped regions at the lowest levels of biological organization. They are not in the business of creating fundamental principles, which they borrow mostly from physics and chemistry. Their spectacular success comes instead from technology invented and applied with creative genius. They render visible, by crystallography, immunology, gene substitution, and other methods, the anatomies and functions of the ultramicro-scopic inhabitants of the cell, which are beyond the range of the unaided human sense. They aim for and can be expected in time to join with researchers in other disciplines of biology to develop the fundamental principles of biological organization.

A large part of the success of molecular biology and cellular biology is due to their importance for medicine. Let me put the matter even more strongly: in public perception and support molecular biology and cellular biology are virtually married to medicine. There is no Nobel Prize in biology, but, expressing what Alfred Nobel thought most important in his 1895 will, there is a Nobel Prize in physiology and medicine. Molecular biology and cellular biology are rich and powerful not so much because they have been successful. They are successful mostly because they have been made rich and powerful. I do not wish to be misunderstood on this point: the investment in these parts of biology by government and the private sector has been worth every penny, and they deserve a still much higher level of sup-port. Their discoveries have unveiled the physicochemical basis of life and set the stage for the eventual elimination of most human disease and genetic disability. Their knowledge has laid part of the foundation of biology at the higher levels of organization.

The second fundamental law of biology is that all biological processes, and all the differences that distinguish species, have evolved by natural selection. Generation by generation, rare and random changes occur in the DNA code. When these muta-tions enable individuals carrying them to leave more offspring in the next generation, the species as a whole changes into the mutant form. The species has thus evolved by natural selection.

When a species changes substantially from its original state, it can be said to have evolved into a new species. When different strains of the same species diverge sufficiently from one another by the accumulation of successful mutations that fit them to different niches, the mother species can be said to have multiplied into daughter species. Charles Darwin, without knowing many of the details, including the existence of genes, nevertheless was able to capture the idea of evolution by natural selection with remarkable clarity and foresight. In the fourth chapter of *On the Origin of Species* the master naturalist encapsulated the idea in one rolling Victorian sentence:

> It may be said that natural selection is daily and hourly scrutinising, throughout the world, every variation, even the slightest; rejecting that which is bad, preserving and adding up all that is good; silently and insensibly working, whenever and wherever opportunity offers, at the improvement of each organic being in relation to its organic and inorganic conditions of life.
>
> *On the Origin of Species,* First Edition (1859)

Thus beyond molecular and cellular biology there remains the rest of biology, comprising the upper levels of the first dimension (organisms to ecosystem) plus almost all of the second dimension (biodiversity) and third dimension (evolutionary biology). Because these domains of investigation began in the eighteenth and nineteenth centuries, they may seem antiquated and declining. The opposite is true. They are a large part of the future of science. As biology matures and unifies, the second and third dimensions will with the upper reach of the first dimension come to overshadow molecular and cellular biology.

The unfolding of the two laws, the physicochemical basis of life and the evolution by natural selection of all known forms of life, defines modern biology. How much, then, of the real living world has biology learned? When the three dimensions—hierarchy, diversity, and history—are taken into full account, it has to be admitted that only an infinitesimal part is known. I would guess that existing biology is under one-millionth of what will eventually be known. That is a very long way to go, but with every added datum and every improvement in technology, additional steps are taken. In the course of this journey biology will continue to progress toward unification. Its leading researchers are coming increasingly to agree that the future of biology depends on interdisciplinary studies within and beyond biology. In time, and the sooner the better, we will be able to travel across the three dimensions without restriction.

13

Exploration of a Little-Known Planet

IN THE LONG JOURNEY AHEAD, biology in general and biodiversity studies in particular need a map. If you wonder, Pastor, what such a prerequisite has to do with the Creation, I have to tell you that we don't know what is happening to most of the rest of life, because we don't even know what it is. Humanity doesn't need a moon base or a manned trip to Mars. We need an expedition to planet Earth, where probably fewer than 10 percent of the life forms are known to science, and fewer than 1 percent of those have been studied beyond a simple anatomical description and a few notes on natural history.

Think about it: if our robotic rovers on Mars discovered life there and sent back accounts of an estimated 10 percent of the species, the American people would gladly spend billions of dollars to find and classify the remaining 90 percent. In sharp contrast, the amount spent on systematics in the United States, from all private and governmental sources was, in 2000, the last year an estimate was made, between 150 and 200 million dollars. That amount was distributed to about 3,000 systematists in this country, out of probably more than 500,000 professionals who can be classified as scientists of all disciplines. To say that humanity has been slow to explore the home planet is an understatement.

The situation in global biodiversity can be recapitulated from my earlier chapters very briefly as follows. Despite the slow pace of exploration, biologists in the past two or three decades have found that Earth's biodiversity is far richer than previously imagined. This diversity is disappearing at an accelerating rate, from habitat destruction, including habitat destruction now underway from climate warming, plus the spread of invasive species, pollution, and overharvesting. If these human-caused forces are left unabated, we could lose as many as half the species of plants and animals on Earth by the end of the century.

Over geological time, and if averaged over many taxonomic groups, species went extinct at the rate of one species per million species per year; and new species were born at the same rate, one species per million species per year. The ongoing rate of species extinction and commitment to early extinction, measured to the nearest power of ten, is now 100 times the rate at which new species are being born, as most conservatively estimated. It is expected to rise to 1,000 or higher, as the last remnants of many ecosystems are wiped out and many species now present on the edge of extinction disappear with them. Biologists who work most closely on biodiversity agree that we are in the beginning of the largest spasms of extinction since the end of the Cretaceous Period, 65 million years ago. In each of the five major prehuman spasms during the past 400 million years, it took about 10 million years for evolution to restore the full amount of biodiversity lost. These estimates are based on the best-known groups, such as mammals, flowering plants, and a few shelled invertebrates such as mollusks. Our ignorance of biodiversity is such that we are losing a large part of it before we even know it existed.

The following figures show how little we have progressed in exploring Earth. The number of species of organisms discovered to date, comprising all known plants, animals, and microorganisms, lies somewhere between 1.5 and 1.8 million. Estimates of the true number, including those discovered plus those still unknown, vacillate wildly according to the method used, ranging (in the 1995 *Global Biodiversity Assessment*) from 3.6 million at the low end to 112.0 million at the high end. Even figures for the relatively well-studied vertebrates are spongy. Estimates for the fish species of the world have varied from 15,000 to 40,000.

The 100-million-plus figure, if it is ever reached, will come mostly from the diversity of invisible organisms. Bacteria and bacteria-like microbes called archaea are the dark matter of Earth's living universe. As of late 2002, 6,288 species of bacteria had been discovered and catalogued. But that many species can be found among the 10 billion bacteria living in only one gram of fertile soil; and an astonishing 4 million species have been estimated to exist in a ton of soil. At least 700 bacterial species thrive as symbionts in the human mouth. They are adapted to life over the vast (for bacteria) plains and canyons of our teeth and tongue, where they are believed to contribute to oral health by excluding disease-causing bacteria. It may seem strange to think of humans in collusion with bacteria, but the truth viewed another way is even stranger: each person's entire body harbors more bacterial cells than human cells. If biological classification were based on a preponderance of cells, a human being would be classified as a bacterial ecosystem.

Other examples exist of the prodigious nature of invisible life. Deep below our feet, and extending for at least two miles down, is another and in some respects far greater world: vast, unexplored populations of bacteria and microscopic fungi, collectively called the SLIMES (subterranean lithoautotrophic microbial ecosystems). The inhabitants may collectively outweigh all of the living matter on the surface of the planet. They depend not on solar energy or organic matter drawn from Earth's surface but on independently ("autotrophically") derived sources of chemical energy in the dissolved minerals that surround them (hence *lithos,* stone). If somehow Earth's surface were burned to a crisp, the life below would likely persist. Then someday, perhaps a billion years into the future, it might evolve new forms of life that could repopulate the surface. The discovery of the SLIMES has given added hope to scientists that life will be found on the bitter cold and powder-dry planet of Mars—not at the surface but far beneath it, at the level of liquid water.

So: we are but one of many species on a little-known planet. Nearly 250 years ago Carolus Linnaeus introduced the practice of giving each species a two-part Latinized name, thus *Homo sapiens* for humanity. He advocated the complete exploration of life on Earth. For the adventure of exploring a little-known planet, and for our own security, it will be wise to press on to finish the great enterprise Linnaeus began. The effort to accomplish a full accounting would be a scientific moon shot, the equivalent of the Human Genome Project, which mapped the letters of almost the entire genetic code of the human species.

To see the potential of this enterprise, imagine an Encyclopedia of Life, with an electronic page for each species of organism on Earth, available everywhere by single access on command. The page contains the scientific name of the species, a pictorial or genomic presentation of the primary type specimen on which its name is based, and a summary of its diagnostic traits. The page opens out directly and through links to other databases. It comprises a summary of everything known about the species' genetic code, biochemistry, geographical distribution, phylogenetic position, habitat, ecological traits, and, not least, its practical importance for humanity.

The page is indefinitely expansible, and its contents are continuously peer reviewed and updated with new information. All the pages together form the encyclopedia, the content of which is nothing less than the totality of comparative biology.

There are compelling reasons to build such an Encyclopedia of Life. Not least is the power it will provide for expanding biology as a whole. As the census of species on Earth comes closer to completion, and as their individual pages fill out to address all levels of biological organization from gene to ecosystem, new classes of phenomena will come to light at an accelerating rate. Their importance cannot be imagined from our present meager knowledge about the biosphere and the species comprising it. Who can guess what the mycoplasmas, collembolans, tardigrades, and other diverse and still largely unknown groups will teach us? As the species coverage grows, gaps in our biological knowledge will stand out like blank spaces on maps. They will become destinations toward which researchers gravitate.

For the first time, the species of entire ecosystems can be censused in full. Unknown microorganisms and the smallest invertebrates, which still comprise most species yet lack even a name, will be revealed. Only with such encyclopedic knowledge can

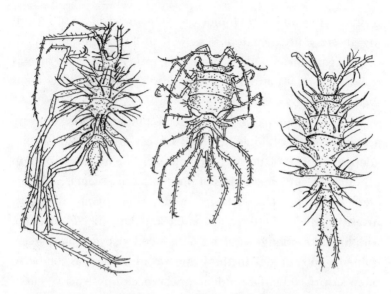

Three species of *Dendrotion* isopod crustaceans from the deep North Atlantic. (From Robert Y. George, "Janirellildae and Dendrotionidae [Crustacea: Isopoda: Asellota] from Bathyal and Abyssal Depths off North Carolina and Their Evolution," *Travaux du Muséum National d'Histoire Naturelle "Griore Antipa"* 47 [2004]: 43–73.)

ecology mature as a science and acquire predictive power species by species and, from that fine-grained information, the same capability for individual ecosystems.

As one practical result, the human impact on the living environment could be assessed in far more reliable detail than is now possible. Today, for example, we base estimates of species extinction on data from a scattering of taxonomically best-known groups, including the flowering plants, land and freshwater vertebrates, and a few invertebrates such as butterflies and mollusks. These taxa contain only about a quarter of the known species on Earth, and almost certainly a much smaller fraction of those still unknown. Tomorrow, other invertebrates, including insects and nematodes, as well as fungi and nearly all microorganisms, together comprising most species on Earth as well as essential pathways of the world's energy and materials cycles, can also be assessed.

The Encyclopedia of Life will serve human welfare in every activity of practical biology. The discovery of wild plant species adaptable for agriculture, new genes for enhancement of crop productivity, and new classes of pharmaceuticals can be accelerated. The outbreak of pathogens and harmful plant and animal invasives will be better anticipated and halted. Never again, with fuller knowledge of such extent, need we overlook so many golden opportunities in the living world around us, or be so often surprised by the sudden appearance of destructive aliens that spring from it.

An Encyclopedia of Life is logically inevitable if for no other reason than that the consolidation of biological knowledge is urgently needed. In its earliest stages, already emerging, it forms a matrix within which comparative studies are rapidly organized. The process will quicken yet more as traditional taxonomic procedures, still mostly dependent on repeated examinations of authenticated specimens and print literature, are replaced by high-resolution digital photography, nucleic acid sequencing, and internet publication. With further documentation organized into the species pages, new lines of research will open at a faster pace. Model species for laboratory and field research can be more easily found—true to the principle that for every problem in biology there exists a species ideal for its solution.

A growing, single-access, species-structured encyclopedia will ease navigation through the already immense biological databases. Aided by computer search engines, we can summon patterns whose detection would otherwise demand overwhelming amounts of effort and time. Principles and theory can be built, deconstructed, and rebuilt with an unprecedented power and transparency.

Ultimately, and at the deepest level, the Encyclopedia of Life is destined, I believe, to transform the very nature of biology, because biology is primarily a descriptive science. Although it depends upon a solid base of physics and chemistry for its functional explanations, and the theory of natural selection for its evolutionary explanations, it is defined uniquely by the particularity of its elements. Each species is a small universe in itself, from its genetic code to its anatomy, behavior, life cycle, and environmental role, and a self-perpetuating system created during an almost unimaginably complicated evolutionary history. Each species merits careers of scientific study and celebration by historians and poets. Nothing of the kind can be said for each proton or hydrogen atom. That, in a nutshell, Pastor, is the compelling moral argument from science for saving the Creation.

IV
TEACHING
THE CREATION

THE ONLY WAY TO SAVE THE DIVERSITY

OF LIFE AND COME TO PEACE WITH NATURE

IS THROUGH A WIDELY SHARED

KNOWLEDGE OF BIOLOGY AND WHAT

THE FINDINGS OF THAT SCIENCE IMPLY

FOR THE HUMAN CONDITION.

14

How to Learn Biology and How to Teach It

THE BASIC INGREDIENT for a love of learning is the same as for romantic love, or love of country, or of God: passion for a particular subject. Knowledge accompanied by pleasurable emotion stays with us. It jumps to the surface and, when summoned, triggers other memory linkages to create metaphor, the cutting edge of creative thought. Rote learning, in contrast, fades quickly into a jumble of words, facts, and anecdotes. The Holy Grail of liberal education is the formula by which passion can be systematically expanded for both science and the humanities, hence for the best in culture.

I cannot define such passion in a few words, because it exists in a multiplicity of unpredictable forms. But I can illustrate it with some confidence, as can most others, from personal experience. Out of my student years at the University of Alabama, I remember in most vivid and resonant detail what I was taught by just three teachers. It has been a long enough interval, over fifty years, for their gift to have passed the test of time.

Septima Smith, a fifty-something spinster, as they called older unmarried ladies in those days, taught medical parasitology with the intensity of a medical school drillmaster. Her intellectual world was the bestiary of microbes and small worms and other invertebrate animals responsible for diseases rampant in rural Alabama. This she insisted every student learn with exactitude and thoroughness. As a sophomore, I was set to scanning my own blood smears and preparations of my own feces (negative, thank heavens, despite all my adolescent excursions into rural Alabama) and using laboratory specimens to trace the life cycles of key pathogenic species. Parasitology was not just a college course with Septima; it was a way of life, and it could have easily been a profession for me had I chosen to continue. Because she cared, I cared. Because Septima Smith expected excellence, she got it. To this day I remember most of the content of the course, and for decades after my studies with Professor Smith I still occasionally used my own drawings of the malarial parasite life cycle in my lectures at Harvard.

Allan Archer was not a teacher, nor did he want to be, which made him all the better one. He was a curator at the Alabama Natural History Museum, located then as now near the center of the university campus. Amiable but shy, he worked alone in a small room at the rear of the museum, reorganizing the spider collection. I started visiting him as an eighteen-year-old, to talk about my studies on ants and to hear his impromptu lectures on the classification of spiders. It was for me an influential bonding with a biologist immersed in an outwardly small yet endlessly intricate part of Earth's biodiversity. Archer was a professional, and he treated me as though I were one also. He gave me self-confidence. He taught me how to talk the talk of a real research scientist. He didn't care about wealth or fame; he cared about the classification and biology of spiders. I didn't understand all the words, but I got the music.

Every student should be fortunate enough to have at least one teacher like Ralph Chermock. He arrived with a new Ph.D. from Cornell, at the beginning of my sophomore year, and took command of my education in biology. As the youngest member of a small group of Chermock disciples (the others were all World War II veterans), I soon was reading and discussing works in the Modern Synthesis of evolutionary theory. Chermock was no abstract dreamer. He believed that evolutionary biology should be built upon a solid bedrock of natural history acquired in the field. "You're not a real biologist until you know the names of ten thousand species." Yes! That is what I yearned to hear: a high goal clearly defined by a charismatic leader. The fauna and flora of Alabama at that time were still poorly known compared with those in the rest of the country. With Chermock's encouragement our cadre of zealots set out on field trips to every corner of the state, from faraway Red Rock Junction to Clayhatchee and on down to Bayou La Batre, and points in between, from the Appalachian foothills to the Mobile-Tensaw floodplain forest, and not least and repeatedly down into the intricate cave systems still mostly unexplored. We collected specimens, specimens, and more specimens, mainly amphibians and reptiles, but also ants and beetles. During three years of such expeditions, we talked natural history and evolutionary biology as we saw the phenomena with our own eyes. We reported to Chermock. Almost unconsciously, we became real, practicing scientists—in fact, our preserved animals and our data are still used. I'm not sure whether any one of us ever mastered ten thousand names of species; like most other people, I tend to forget old ones as I learn new ones. But the subject we inhabited in the field and the joy we drew from the hands-on training entered our bones and shaped our souls. Every one of us subsequently became a professor of biology. More than fifty years later we still refer to ourselves as the Chermockians.

Education in biology is important not just for the welfare of humanity but for the survival of the rest of life. Every conservationist with whom I have discussed the subject agrees that the general indifference of people to the living world is the failure of introductory education in biology. The shortfall has been worsened by the common misperception that "rigorously scientific" biology means molecular biology, neurobiology, and biomedical research; it does not mean evolutionary or environmental research. But, as I have urged, half of biology now, and probably more than half in the future, lies in the study of biodiversity and the living environment. Within this domain exists much of the unique intellectual content of biology, and a part also of immediate relevance and potential interest to the public.

The breadth of biology offers entry to a liberal education, which sets out to develop human beings who know not just facts but concepts, understand how to learn, and are able and motivated to think for themselves.

How can biology best be made part of a liberal education? I believe I can provide an answer. For most of the forty-one years I served on the Harvard faculty, I was privileged to teach beginning biology, chiefly to nonmajors as part of what at least passed as the liberal arts program. My focus was on the level of organisms and ecosystems. With my students I also explored fully the evolutionary process. The effort was at any rate a popular success: my student ratings were high, and I received both of the college-wide teaching awards. I believe that the principles I learned over the years about teaching, both by listening to great lecturers at Harvard and by my own trial and error, can apply to both undergraduate and graduate programs everywhere, as well as to secondary school courses at a more advanced level. The relevance of the principles has been confirmed during lecture stints and discussions I have held in many universities and liberal arts colleges across the United States and abroad.

The first principle is:

Teach top-down. If I learned anything in four decades of experience, it is that the best way to transmit knowledge and stimulate thought is to teach each subject from the general to the specific. Address a large question of the kind already interesting to the students and relevant to their lives, then peel off layers of causation as currently understood, and in growing technical and philosophically disputatious detail, in order to teach and provoke. Explain, for example, aging and death as best can be done with knowledge of evolution, genetics, and physiology, then explore the consequences in demography, public policy, and philosophy. Finally, proceed laterally, if wished, into the consequences of the phenomenon to history, religion, ethics, and the creative arts. Do *not* teach from the bottom up, with an introduction such as "First, we'll learn some of this, and some of that, and we'll combine the knowledge later to build the bigger picture." Don't paint the picture in pointillist dabs to easily bored students. Instead, put it up whole as quickly as possible, and show why it matters to them and will matter for a lifetime. Then dissect the whole down to the foundations.

Take, for example, sex. Not the anatomy and practice thereof, nor physiology, fertility, or birth control. Instead ask, Why is there sex in the first place? How does the biologist view the matter, as opposed, say, to the philosopher, theologian, or novelist? Why don't people—women, to be precise—just practice parthenogenesis by growing embryos from unfertilized eggs? This no-sex practice is widespread in the animal kingdom. Why do there have to be males and sperm anyway? The answers to these unfamiliar questions, if one doesn't just stop at Adam, Eve, and God's will in searching for the ultimate cause, leads to the issue of genetic variety. Having two genetic codes gives each person flexibility to deal with a constantly changing environment. Over much of Sub-Saharan Africa, to take the classic case, having a gene for sickle-cell anemia from one parent protects you from malignant malaria, while having a matching normal gene reduces the effect of the sickle-cell gene enough to keep anemia from killing you. The result is that the sickle-cell trait is widespread wherever malignant malaria is common, but never replaces the normal gene.

In general, having two genetic codes also allows the parents to create offspring with more genetic variety among them, so that at least one or a few will survive in a constantly changing environment. Yet genetic variety as the ultimate cause of sex is only a theory. How would biologists go about testing this theory? Has it been proved? (In fact, it is strongly supported, but not yet definitively proved, by the evidence.)

By such means provoke the students, give them a new slant, challenge the assumptions and comfortable beliefs they brought with them, turn them into colleagues, propel them on intellectual and spiritual searches of their own. Thereby prepare them properly to enter, as Harvard's commencement blessing intones, the fellowship of educated men and women.

Like science teachers everywhere, I encountered a major

obstacle in math phobia, the pandemic curse of *Homo sapiens* in training. I'm sure that many of the students at Harvard chose to major in the humanities, and face the different, more verbal rigors of that domain, or at least took as little science as possible, because they believed they lacked mathematical ability. The subject matter of science might be fascinating to them—the origin of the universe, the nature of climate change, the evolution of life, and, of course, the meaning of sex—but the required quantitative mode of thinking seemed too daunting.

Math phobes are wrong! Mathematics is just a language, and language is only a habit of thought. Chinese idiograms and mathematical arguments are equally mystifying to the uninitiated, and equally familiar to those who learn them early in life. Once the standard symbols and operations of mathematics are learned and used repeatedly to the point of second nature, scanning an equation is not very different from reading a passage in a book. A text on population genetics can be less mystifying than *Ulysses,* and vastly easier than the untranslated *Beowulf.*

Those who have avoided the language of mathematics are best led into it with a top-down approach to some important and interesting problem in real life. Here is one of my favorite examples. Few things are of more concern to people than hereditary diseases or proneness to disease. Defective genes occur in all human populations and manifest themselves in virtually every category of illness, mild to fatal, from spontaneous abortion and infant mortality to hundreds of child and adult disorders. Hemophilia, sickle-cell anemia, cystic fibrosis, Huntington's chorea, and individual forms of color blindness are among the most familiar. How common are these genes, and how common are the symptoms they cause?

Bear with me now for the next two paragraphs while I walk through the explanation given my annual congregations of Harvard math phobes. Once a student has learned the elementary principles of Mendelian heredity, which are actually mathematical formulas without the abstract mathematical notations, he is ready for the Hardy-Weinberg equation, a cornerstone of population genetics and evolutionary theory. The equation says, Consider that each person has two chromosomes of the same kind and that at any given position on the chromosomes is a gene that can differ (or not differ) going from one of these chromosomes to the second of the chromosomes. In a population of people, count the number of genes of each kind (remember, there are two genes on each chromosome position per person, one from each parent, and as a result twice as many genes at that position as there are people). Take the percentage of genes of the first kind on the selected position, say 80 percent (a frequency of 0.8) and 20 percent (a frequency of 0.2) of the second kind. The Hardy-Weinberg equation states that the frequency of organisms (in this case, people) in the population carrying two of the first kind of gene at the selected position is the square of the frequency of that gene, or $0.8 \times 0.8 = 0.64$; and the frequency of organisms in the population carrying two of the second gene is the square of the frequency of that kind of gene, or $0.2 \times 0.2 = 0.04$. Finally, the percentage of organisms in the population with one of each kind of gene is the multiple of the two gene frequencies times 2; in this example, $0.8 \times 0.2 \times 2 = 0.32$. The three frequencies must add up to 1.0, or 100 percent, and they do: $0.64 + 0.04 + 0.32 = 1.0$.

That's it. That's all. Now you can express the principle as a mathematical equation: $p^2 + 2pq + q^2 = 1.0$. Converted to numbers, the equation is $(0.8 \times 0.8) + (2 \times 0.8 \times 0.2) + (0.2 \times 0.2) = 1.0$. You can also derive the Hardy-Weinberg equation from first principles of Mendelian heredity, much the way Godfrey H. Hardy and Wilhelm Weinberg did it a century ago, and on the back of an envelope.

Why does the Hardy-Weinberg equation matter? Start with ordinary genes that can be detected at a glance, many of which are recessive (their effects blocked by the presence of the dominant gene) but expressed when in double dose. Examples that the students, sitting in class, can check on their own person include ear lobe attached to head versus earlobe hanging free, inability to roll the tongue into a tube, widow's peak at the front margin of the hair, hitchhiker's thumb (ability to bend it far back). From this, we estimate immediately the gene frequencies in the population, as well as the frequency of individuals with double doses and those with half doses of the dominant gene. Now, a teacher need only point out that while attached earlobes and widow's peaks carry no apparent liability, the same Hardy-Weinberg equation holds for genes that cause disease. The principles are therefore an important part of modern medicine. Almost every student knows of someone, often a relative, carrying such defective genes.

The second principle is:

Reach outside biology. The ongoing explosive growth of knowledge, especially in the sciences, has resulted in a convergence of disciplines and created the reality, not just the rhetoric, of interdisciplinary studies. Biology, for example, is today a swiftly evolving kaleidoscope of hybrid subdisciplines. Professional journals and college curricula teem with such names as molecular genetics, neuroendocrinology, behavioral ecology, and sociobiology.

Biology has also expanded to the borders of the social sciences and humanities, and they to it. As a consequence, what was once perceived as an epistemological divide between the great branches of learning is now emerging from the academic fog as something far different and much more interesting: a wide middle domain of mostly unexplored phenomena open to a cooperative approach from both sides of the former divide. Already disciplines from one side of this middle domain—for example, neuroscience and evolutionary biology—have connected with their closest neighbors, psychology and anthropology, on the other side.

The middle domain is a region of exceptionally rapid intellectual advance. It, moreover, addresses issues in which students (and the rest of us) are most interested: the nature and origin of life, the meaning of sex, the basis of human nature, the origin and evolution of life, why we must die, the origins of religion and ethics, the causes of aesthetic response, the role of environment in human genetic and cultural evolution, and more.

The third principle is:

Focus on problem solving. If the top-down concept of presentation works, and it does, and further given the convergence and blending of disciplines, the best approach to general education in the future would seem to be less discipline-oriented and more problem-oriented. The problem (or big issue) addressed singly within a given course top-down could be of the following kind: the nature and consequences of human nature, the basis of moral reasoning, or the crisis of global freshwater supply and its solution. Such an approach would require some breadth on the part of the instructor, or at least team-teaching by a group of complementary experts.

There is, in my opinion, an inevitability to the unity of knowledge. It reflects real life. The trajectory of world events suggests that educated people should be far better able than before to address the great issues courageously and analytically by undertaking a traverse of disciplines. We are into the age of synthesis, with a real empirical bite to it. Therefore, *Sapere aude.* Dare to think on your own.

The fourth principle is:

Cut deep and travel far. By the sophomore year all college students should have begun some strategic thinking about their own education. The best pattern to follow is T-shaped. The vertical shaft represents the drive deep into a specialty, and the horizontal bar the breadth of experience obtained from a liberal education. The specialization is for a trade or preparation for graduate school. The liberal arts are for flexibility and maturity of intellect. Of course, this is already the combination intended by most universities and four-year colleges. Students are expected to select a "major" or "concentration" such as English, economics, or biology by the second year and in addition to take elective courses scattered across the intellectual landscape. But most students have to be persuaded that such is the best strategy for them.

For future biologists, I offer the same advice I gave hundreds of students at Harvard regardless of their career plans. As soon as you feel comfortable doing so, choose a part of biology to which you then commit yourself, and treat the rest of biology as general education. Trust your instincts, press on into molecular biology, or behavioral biology, or ecology, or some other discipline or combination of disciplines within the biological sciences broadly defined. Knock around a bit to locate more precisely your future intellectual home base.

Although, as expected, a majority of students assigned to me as biology concentrators were aiming for medical school, a quarter or more wanted to be field biologists. They made this choice even though opportunities for careers were consistently few. I never wavered in my advice to these would-be naturalists: follow your heart.

The fifth and final principle is:

Commit yourself. Returning to passion as the driver of learning, a teacher's dedication is most effective when expressed through both the art of teaching and the demonstrated love of the subject for its own sake. Secondary school and college students seek their personal identity, but they also yearn for a cause larger than themselves. By some means they will acquire both these marks of maturity, whether base or noble. In transit they need mentors to trust, heroes to emulate, and accomplishments that are real and enduring.

I will next argue that Nature is a theater for which such mental development is inherently suited.

15

How to Raise a Naturalist

THE ASCENT TO NATURE begins in childhood, and the science of biology is therefore ideally introduced in the earliest years. Every child is a beginning explorer naturalist. Hunter, gatherer, scout, treasure seeker, geographer, discoverer of new worlds, all these are present at the child's inner core, rudimentary perhaps but straining for expression. Through time immemorial children were reared in intimate contact with natural environments. The survival of their tribe depended on a close, tactile knowledge of wild plants and animals.

Then, after millions of years of such existence, the agricultural revolution abruptly removed most people from the habitats in which their ancestors had evolved. It allowed them to multiply to higher population densities, but at the price of chaining them to much simpler surroundings. They came to depend on a drastically reduced number of plant and animal species, which could be cultivated only in a biologically pauperized environment by repetitive labor. As the still larger populations supported by agricultural surpluses emigrated into villages and cities, people drifted still farther from the ancestral environment. Today, most of humanity dwells in an artifactual world. The cradle and original home of our species has been largely forgotten.

The ancestral instincts nevertheless still live within us. They are expressed in art, myth, and religion, in gardens and parks, in the strange (when you think about it) sports of hunting and fishing. Americans spend more time in zoos than at professional sports events, and more time yet again in the increasingly crowded wildlands of national parks. Recreation in national forests and nature reserves—the parts that remain uncut, that is—generate substantial wealth, over $20 billion annually, for example, to the American gross domestic product. Wild nature saturates television and movies in the industrialized world. A touchstone of personal wealth is the second home, typically in pastoral or natural environments. It serves as a retreat for peace of mind and a return to something otherwise lost but not forgotten. Bird-watching, or birding, as the cognoscenti prefer to call it, has become a major hobby and a robust industry.

To be a naturalist is not just an activity but an honorable state of mind. Those who have expressed its value and protected living Nature are among America's heroes: John James Audubon, Henry David Thoreau, John Muir, Theodore Roosevelt, William Beebe, Aldo Leopold, Rachel Carson, Roger Tory Peterson. Cultures around the world still living close to Nature value talent in natural history. Those dependent on artisanal hunting and fishing, and on sustenance agriculture, stake their lives on knowing it well. The cognitive psychologist Howard Gardner has defined such ability as one of the eight major categories of intelligence:

A naturalist demonstrates expertise in the recognition and classification of the numerous species—the flora and fauna—of his or her environment. Every culture prizes people who not only can recognize members of a species that are especially valuable or notably dangerous but also can appropriately categorize new or unfamiliar organisms. In cultures without formal science, the naturalist is the person most skilled in applying the accepted "folk taxonomies"; in culture with scientific orientation, the naturalist is a biologist who recognizes and categorizes specimens in forms of accepted formal taxonomies.

The cognitive skills of the talented naturalist play out in many other ways, including the practical activities of industrialized societies. "The young child who can readily discriminate among plants or birds or dinosaurs," Gardner observes, "is drawing on the same skills (or intelligence) when she classifies sneakers, cars, sound systems, or marbles"; and "it is possible that the pattern-recognizing talents of artists, poets, social scientists, and natural scientists are all built on the fundamental perceptual skills of naturalist intelligence."

I argued earlier that biophilia, the inborn attraction to the natural world, has provided individuals and tribes an adaptive edge throughout evolutionary history. Now natural history is coming back to biology in a way that will expand its base into a more human-oriented and humane science.

How best to cultivate a naturalist's intelligence in every child? And how to promote excellence in those who prove talented in natural history? These questions having received very little attention from research psychologists, I will presume to draw again on my personal experience and what I have learned by talking with parents and teachers, as well as children, over a period of many years.

A child's mind opens to living Nature early. If stimulated, it then unfolds in stages that strengthen the bond to nonhuman life. The brain is programmed for what psychologists call prepared learning: we remember with ease and pleasure some experiences. In contrast, we are counterprepared to avoid learning, or else to learn and then avoid, other experiences. For example, flowers and butterflies yes, spiders and snakes no.

The rationale from evolutionary biology concerning such biased learning is straightforward: the cues that signal the healthful, productive part of the environment result in genetically swift positive reinforcement and do not need to be taught or repeated; those that signal danger result in similarly swift negative reinforcement.

I have several time-tested suggestions for parents and teachers, including religious leaders, who wish to cultivate the naturalist's capability in a child. Start early; he is ready. Open doors to Nature, but don't push him through. Think of the child as a hunter-gatherer. Provide opportunities to explore the outdoors and its surrogates in zoo and museum exhibits. Let the child search, alone or in small, like-minded groups. Let him disturb nature a bit, on his own and without coaching. Provide field guides, binoculars, and even microscopes, at home if possible and at least at school. Encourage and praise his initiative. With adolescence, allow him to undertake adventures with others, to wild areas and foreign countries as opportunity and finances allow. Let him learn all things at his own pace. At the end of this process he may choose a career in law, marketing, or the military, but he will be a naturalist all his life, and thank you for it.

I hope the foregoing recommendations make it clear that becoming a naturalist is not like studying algebra or learning a foreign language. It would be a mistake to introduce a child to Nature by a walk through a park or arboretum, with labels naming the species of trees and shrubs. The child is a *savage*, in the best meaning of this word. He needs to thrill to the excitement of personal discovery, to mess around a lot and learn as much as possible on his own.

Try this. Buy him a small compound microscope; they are now available at no greater cost than a skateboard or airfare to Disney World. Suggest that he look at drops of pond water, sampled with an eyedropper from aquatic plants or algae. Don't tell him what to expect, only that it will be unlike anything he has ever experienced. He will see what astonished Robert Hooke, Antony van Leeuwenhoek, and Jan Swammerdam, the first microscopists of the seventeenth century: a miniature Jurassic Park, inhabited by translucent shape-changing rotifers that snake their way through the detritus, settling and opening out their hairlike cilia on the head to create circular water currents; protozoans darting and spinning through the water and bumping into obstacles like drunken drivers; crystalline diatoms; and more, almost infinitely more.

I had this experience at the age of eight. My parents gave me a microscope. I don't recall why, but no matter. I then found my own little world, completely wild and unconstrained, no plastic, no teacher, no books, no anything predictable. At first I did not know the names of the water-drop denizens or what they were doing. But neither did the pioneer microscopists. Like them, I graduated to looking at butterfly scales and other miscellaneous objects. I never thought of what I was doing in such a way, but it was pure science. As true as could be of any child so engaged, I was kin to Leeuwenhoek, who said that his work "was not pursued in order to gain the praise I now enjoy, but chiefly from a craving after knowledge, which I notice resides in me more that most other men."

The thirst for knowledge can be heightened by repeating the archetypes that rule the developing mind. At eight to twelve years of age, many children establish secret places, ideally caves or abandoned buildings, but in fact any out-of-the-way spot that offers privacy. A shelter can be built from saplings (which I used, although they turned out to be poison oak!), scraps of lumber, abandoned cinder blocks, or other makeshift materials. A tree house is ideal, because it offers maximum privacy and protection. Woodland, even a small fragment of second growth, is a logical choice of habitat. In the secret place the child, perhaps along with a couple of friends, collects magazines, reads and talks a lot, and monitors the surrounding terrain.

Children are born treasure hunters and collectors. Given any access to natural environments, they are likely to start searching for minerals ("gems"), specimens of butterflies and other insects, and small live animals of any kind. Encourage this activity. Do not allow yourself to be squeamish. Pet toads, snakes (nonpoisonous), and minnows are just fine. Testing the outer limits of my parents' tolerance, since I already brought home snakes, I kept and bred black widow spiders, feeding them live flies and cockroaches. Ant colonies housed in artificial nests ("ant farms") are potent in every way: the workers are in a flurry of activity day and night; they quickly convert a small pile of earth into a home, from which they lay down invisible odor trails to newly discovered food. Ants are as relaxing as fish in an aquarium and make excellent science projects for school.

For maximum impact in a short period of time, take a child to the seashore and challenge him to make a collection of creatures he finds on his own. In settled areas and heavily used beaches, use a digital camera for all but the smallest animals, and otherwise collect everything live for return to the sea. Along sandy beaches, legions of little insects, crustaceans, and bivalve mollusks lurk in drifts of stranded seaweed; mysterious dead animals or their fragments wash ashore from deeper water. In the tide pools of rocky shores dwell a seemingly endless variety of small crustaceans, snails, sea anemones, sea urchins, starfish, and other, less familiar inhabitants of the shallow marine environment. After a while, open a field guide and help the child put names on the discoveries. And if a small compound microscope is available as well, encourage him to sample drops of water drawn from around the algae and rock surfaces. Thus add another and even richer world of biodiversity.

Adventure with a different feel to it awaits the child who joins a group of birders. As an adult I thrill, even myopic entomologist that I am, at the sight of eagles, cranes, and ibises. Recently I sat in a skiff on Mississippi's Pascagoula River, transfixed by a dozen swallow-tailed kites that wheeled overhead and swooped to take sips of water from the river.

It is among birders, all of them naturalists and adventurers, that the child can find role models. There are a few eccentric loners in their ranks, but also physicians, ministers, plumbers, business executives, military officers, engineers, and in fact members of virtually every trade and profession. They are united in a common focus. At least while in the field, they are among the most congenial and enthusiastic people I have ever known.

Take the child to the zoo, with a purpose. Don't wander through the exhibits passively, but instead pick a kind of animal for close-up study. Reptiles are popular, and of course large mammals always, but so are the littlest of the creatures on display. For years one of the parts of the National Zoological Park in Washington, D.C., that has attracted the most visitors has been the insect collection. Among the exhibits there, since their inception in 1987, the most popular has been the Soil Table, a long trough filled with soil and leaf litter from a nearby woods. Visitors, mostly boys and girls, explore this miniature terrain to glimpse the myriad insects and other small invertebrates living there. They are allowed to comb and pick through the material like entomologists in the field to expose and identify the inhabitants.

Visit an aquarium for similarly high impact. People, including children, love sharks almost as much as they do dinosaurs—and sharks can be seen *alive*. But they are drawn equally to the brilliance of reconstructed coral reefs and the great diversity of life in and around them that can be taken in with one glance. Visit a botanical garden, enter a simulated rainforest and drink in the grandeur it represents. Or study the occasional exhibitions of orchids as you would fine paintings in a gallery. They are the most diverse flowering plants on Earth as well as arguably the most aesthetically pleasing.

From the freedom to explore comes the joy of learning. From knowledge acquired by personal initiative arises the desire for more knowledge. And from mastery of the novel and beautiful world awaiting every child comes self-confidence. The growth of a naturalist is like the growth of a musician or athlete: excellence for the talented, lifelong enjoyment for the rest, benefit for humanity.

16

Citizen Science

WE ARE DRAWING CLOSE to the end of the journey I invited you to take with me. There is still more to becoming a naturalist than personal fulfillment and the conservation of life, although these are surely more than enough. Scientific natural history is also one of the few endeavors in which almost any interested person can make original contributions to science. The data collected go directly into permanent records utilized by researchers in ecology, biogeography, conservation biology, and other specialized fields.

The information from citizen scientists is needed, now more than ever, and it has permanent value. The data will not be treated as redundant or merely confirmatory to knowledge already acquired. There are just too many kinds of organisms and too few professional scientists to study them for anything approaching saturation. I've noted that between 1.5 and 1.8 million species have been described to date, and at least 10 million more await discovery. Even among those known, fewer than 1 percent have been studied in any depth. Their geographical distributions need to be mapped, their habitats recorded, their population size estimated, and their life cycles tracked. How many professional and semiprofessional scientists are available for this research? Addressing the identifications and classification of organisms worldwide are just six thousand experts, of which about half reside in the United States. To move the exploration of Earth's fauna and flora forward, these overworked researchers need more eyes, more boots on the ground, and more fresh ideas.

Just this kind of collaboration between professional and lay researchers has begun to spread around the world. At the cutting edge is the effort to make complete censuses of all forms of life found in selected localities. All-species inventories of this kind have begun to multiply in sites that include ponds and lakes in Denmark and Japan, rainforests in Costa Rica and the Amazon, the Galápagos Islands, and—thanks to legions of devoted naturalists for more than two centuries—virtually the whole of England.

One of the most intense such initiatives in the United States is the one ongoing (as of 2006) in the Great Smoky Mountains National Park, a reserve stretching across the southern Appalachian chain in North Carolina and Tennessee. This all-species project, called ATBI (for All Taxa Biodiversity Inventory), has enlisted experts on different kinds of organisms from all over North America. Assisted by volunteers, and with only a shoestring budget, they have built it into a major enterprise of biological research, as well as teaching center for students at every level from grammar school to Ph.D. and postdoctoral programs.

The mountains of southern Appalachia compose the most ancient range in North America never blanketed by continental glaciers. Their forests are correspondingly the richest in biodiversity. In the upland streams swarm mayflies, stone flies, and other delicate and ephemeral insects, with ancestral lines older than the Age of Reptiles. The highest concentration of salamander species known in the world live in the mountains and foothills, variously brown, yellow, gold and green, black and red, splashed in different patterns. Minnow species found nowhere else change from valley to valley. Legions of tardigrades, slow-moving spore eaters also known as bear animalcules; springtails that can vault the human equivalent of a kilometer; armored oribatid mites that resemble a cross between a spider and a turtle; entotrophans, japygids, nematode worms, and other tiny invertebrates that only experts can recognize, populate the soil. They are but the tip of diversity, rivaled in species numbers by the fungi and vastly exceeded by the bacteria.

The yield of the Great Smoky Mountains inventory has been impressive. From early days in 1998 to the summer of 2004, a total of 3,314 species of all categories of organisms were added to those previously recorded from the Park, hence to the known composition of Appalachian mountain ecosystems; and 516 species were discovered that were entirely new to science, in other words, never before seen anywhere. Some of these novelties are microscopic and obscure. But not all. Twenty-eight of the new species are crayfish and copepod crustaceans, 25 are beetles, and 72 are butterflies and moths. It should be kept in mind that these discoveries are being made not in a remote Amazon camp but in localities within an easy drive for tens of millions of Americans.

The spirit of the collaborative research is captured in this account by David Wagner, the leader of the Lepidoptera (moth-and-butterfly) team:

At 3:00 pm on 19 July, 2004, we poured out of the Sugarlands Training Room and fanned out to the far reaches of the Park. Our odd collecting gear, sheets and traps illuminated with mercury vapor bulbs and blacklights, were set out at more than 40 trapping stations, representing the Park's many elevations, plant communities, and forest types. The night's treasures—moths, legions of the night—were brought to Sugarlands by 8:00 am to be sorted, identified, counted, databased, and vouchered over a non-stop two-day effort. It was a focused full-throttle effort fueled by endless cups of coffee and donuts, and by the time the dust and scales had cleared on Wednesday afternoon, the sleep-deprived 40-member team had recorded and vouchered 795 moth and butterfly species.

DNA samples were taken from 642 of these species for later sequencing. By thus decoding a 700-base-pair section from each specimen's mitochondrial genome, and entering the data in a "Barcodes of Life" website, scientists could identify many of the species collected on later excursions, even from fragments of adult tissue or caterpillars. Because caterpillars are completely different in outward appearance from the adult moths and butterflies into which they metamorphose, and also sampled by the Wagner team, their DNA sequences are needed to record the food plants of the species and complete the tracing of their life cycles.

The barcoding option illustrates how quickly different fields of biology are coming together in citizen-aided surveys. Since the 1990s, advances in technology have been speeding the exploration of biodiversity with equal effectiveness around the world. High-resolution digital photography is added to a computer program which, in a manner similar to that used in medical tomography, creates perfectly focused three-dimensional images of even the smallest insects and other organisms. The images are then electronically transmitted, allowing almost instantaneous sharing of information. Museums and herbariums have begun to photograph and put online images of known plant and animal species, some from authenticated specimens over a century old. On the drawing board is the remote-controlled, robotic examination of specimens, which will allow investigators to manipulate and magnify museum specimens from anywhere on Earth. These procedures will make it far easier to update classifications and to speed field research on biodiversity still more.

The convergence of databases on biodiversity into a few free, single-access, on-command systems has begun to benefit biologists and students dramatically. Question: Do you wish to take a field guide to the butterflies of Argentina on your next excursion to South America? Have one handy for the freshwater fishes of Botswana? The ferns of Sumatra? All the plants and animals of Rock Creek Park? No problem. Within a decade or two it will be possible to assemble a made-to-order field guide of any group, to the degree that it has been explored, living in any part of the world. I have already begun to do so routinely during my field excursions to study ants in the West Indies. It will also be possible, when enough authenticated images of the world's plant and animal species are available, to make up field guides as needed even while in remote field camps.

The next stage in the mapping of Earth's biodiversity is the assembling of the aforementioned Encyclopedia of Life, a program already begun at the U.S. National Museum of Natural History. An electronic page is created for each kind of organism, previously known or newly discovered, into which everything learned about the species is recorded and continuously updated. It is here that students and citizen scientists can make their second important contribution. Scientific natural history, from details of life cycles to natural behavior and the functioning of ecosystems, is an important endeavor for the future of biology. But it is labor-intensive and relatively slow, and in the study of less common species often dependent on chance encounter. Even a professional specialist can hope to make only a limited number of such discoveries in any given year for any particular kind of organism. The collaboration of amateur naturalists improves the process substantially. Consider: one observer may witness a population of butterflies using one kind of larval food plant in Sweden, at the northern limit of the species range, and a second finds the same species on an entirely different food plant in the southernmost population, in Italy. Or, a species of frog may be increasing in Kansas but declining toward extinction in Colorado. A butterfly could be found rare in Fiji but exploding to pest proportions in Samoa. Such are the fine-grained data needed to track the impact of climate change and other trends in ecology.

The engagement of citizen scientists in biodiversity exploration often starts with bioblitzes, which are treasure hunts designed to find and identify the largest number of species possible in one place during a twenty-four-hour period. Experts, professional and amateur alike, gather at an interesting site at an appointed time for a pep talk and invitations to later lunch-

eons and dinners laid on by local residents. Then they fan out
in all directions to locate and identify as many species of their
chosen category of plant or animal as possible. In small groups,
each ideally led by an expert and often including students,
friends, and interested hangers-on, they begin to list birds,
dragonflies, lichens, trees, mosses—any category that has a
capable guide available. Specimens of common species are collected and photographs of rare ones taken. At the end of the
period, all gather to combine and tabulate the results. Bolstered
by food and a pleasing variety of refreshments, the adventurers
exchange notes and war stories: "I think I got what may be a
new species of ground beetle; either that or it has to be some
kind of a tremendous range extension." "Lemme see that; I
must have picked up the same thing. I'll bet it's a recently introduced alien." The most valuable specimens are sent on their
way to museums and herbaria, for use by specialists.

The first bioblitz was, to the best of my knowledge, the one
held out of Walden Pond in Massachusetts on July 4, 1998, and
in addition covered adjacent areas in Concord and Lincoln.
Walden Pond was chosen because it was the site of the cabin
where, in his two-year hermitage, Henry David Thoreau conceived the founding philosophy of American environmentalism.
July 4, 1845, was the date he moved into the cabin. Our event
was called Biodiversity Day. Conceived and organized by Peter
Alden, a local resident and international wildlife tour guide, it
drew over a hundred experts from around New England. I par-
ticipated as sponsor and one of the ant specialists. We aimed for
1,000 species of all kinds of plants and animals and got 1,904—
actually 1,905, if you count the moose that wandered in and out
of Concord Center the next day.

So popular was Biodiversity Day that in the following year the Massachusetts Department of Environmental Affairs extended it in multiple localities to include students from selected school districts. The next year all the districts in the state were engaged in the program.

By 2006, as I write, bioblitzes have been conducted in six other states (Connecticut, Illinois, New York, Pennsylvania, Rhode Island, Virginia), and seventeen other countries (Austria, Belgium, Bolivia, Brazil, China, Colombia, France, Germany, Hungary, Italy, Luxembourg, Netherlands, Norway, Panama, Poland, Switzerland, Tunisia). One of considerable symbolic magnitude was held in New York's Central Park on June 27, 2004, during which, in the words of two Explorers Club participants, Richard C. Wiese and Jeff Stolzer, they together with experts, students, and sundry New Yorkers "crawled through the woods, went diving in a lake, climbed trees, chased butterflies, and reveled in the natural wonder of a beautiful park in their quest to discover new life." The park is indeed beautiful. It is rendered more so through the contrast of its greenery to Manhattan's stone mountains that shadow it and the rivers of humanity that flow around and through it. There is even a hint of wildness: a small tract of undisturbed hardwood forest has

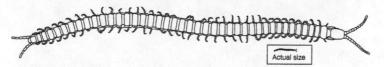

A species of centipede, possibly the world's smallest, new to science and different enough to rank as a new genus, discovered in New York's Central Park in 2002. (From Kefyn M. Catley, American Museum of Natural History.)

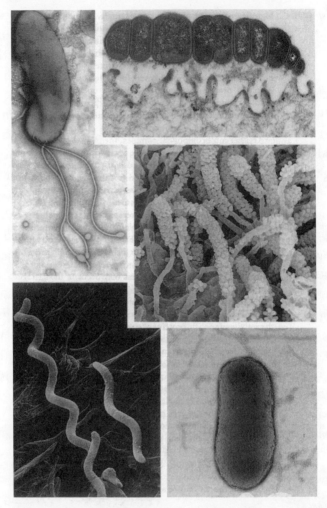

A medley of bacteria. The spiral species at bottom left is a free-living aquatic species. The rest are inhabitants of various parts of the human alimentary tract; at bottom right is *Escherichia coli,* common in polluted water and a key species in molecular biology research. (From Paul Single-ton, *Bacteria in Biology, Biotechnology and Medicine,* 6th ed. [Hoboken, N.J.: John Wiley, 2004], p. 12.)

been left intact near the center. In 2004 a feature new to bioblitzes was added, a dive into the smaller of the two lakes, led by the famous marine underwater explorer Sylvia ("Her Deepness") Earle. Even though Central Park measures only 843 acres, the twenty-four-hour search yielded 836 species of plants and animals.

Now it is the turn of invisible life to be revealed. The reach of the collaborative teams has begun to extend into the almost wholly unknown world of bacteria. The millions of species living in a few thousand kilograms of fertile soil are virtually all unknown to science, anywhere, under any conditions. Through mid-2004 only ninety-two species had been recorded in the Great Smoky Mountains National Park. There are probably that many in a pinch of soil the size of a pencil eraser. The total number in the park could easily be in the tens of millions. New technologies in cell cloning and DNA sequencing have recently made it possible to make a quantum advance in the separation and identification of bacterial species. The methods, already fast, will soon be much faster as well as affordable. Microbiologists agree that in time it will be possible to carry sequencing instruments into the field, along with genomic database software for immediate identifications of species as they are collected.

The portable nature of biodiversity technology also makes it an ideal conduit for transfer of frontline biological research to developing countries. The recently formed Consortium for Biodiversity of the Caribbean is an example of how quickly such an expansion can be accomplished. The consortium includes U.S. institutions such as the Smithsonian Institution and New York Botanical Garden, as well as the Natural History Museum and

Butterflies and moths of the Dominican Republic. (From Biocaribe.org, by permission of Brian D. Farrell.)

National Botanical Garden of the Dominican Republic. The garden, occupying two square kilometers within the otherwise congested capital city of Santo Domingo, is one of the largest such urban reserves in the world. It also contains, very notably, a more than half a square kilometer patch of rare old-growth lowland rainforest. With the support of the consortium, a network of scientists have set out to explore thoroughly the flora and fauna of the Dominican Republic, and more widely the remainder of the West Indies, while making the information electronically available. The effort has a side benefit: just as in industrialized countries, the information technology and biodiversity science employed here can be directly introduced into local educational curricula, all the way from grammar school through college.

I was swept into this effort when I was in my early seventies, and thought my serious fieldwork had ended. I led a team that worked from the dry scrubland forest of the east coast into the surviving remnants of mountain rainforests and then higher, to the pine-bunchgrass savanna at 2,440 meters (8,000 feet) in the central cordillera. I felt the same exhilaration I had experienced in Cuba and the South Pacific fifty years previously. At the most fundamental level nothing in my enthusiasm for biodiversity research had changed, except for a grand goal that now seemed within reach.

Given the richness of tropical biodiversity, and the slow pace of earlier explorations, the research in the Dominican Republic has begun to pay off quickly. Brian Farrell, the Harvard University entomologist who conceived the consortium and has led its overall effort, recently described one of the first practical applications of the new wave. Among the collections made by a group

of Harvard and Dominican Republic students were two unfamiliar black-and-white butterflies:

> These specimens would soon prove to be remarkable dis-
> coveries—not just the very first records for the Dominican
> Republic, but also the very first documentation of this par-
> ticular species, the lime or checkered swallowtail (known
> technically as *Papilio demoleus*) in the Western Hemisphere!
> The lime swallowtails of the Old World tropics are swift-
> flying butterflies whose caterpillars defoliate young lime
> trees, orange and other citrus crops throughout southeast
> Asia, India, and neighboring regions. They can com-
> pletely strip the leaves from young nursery trees and cause
> many millions of dollars in damage annually. This species
> therefore poses a possibly significant threat to citrus indus-
> tries in the Dominican Republic.

The all-species inventories of the Great Smoky Mountains
National Park and West Indies are among dozens of such enter-
prises that have arisen around the world to accelerate the exploration of Earth's biodiversity. Using new biological and
information technologies, they range in scope from state or
even municipal censuses, the latter including the Chicago
Wilderness and Boston Harbor Islands initiatives, to programs
that are continental or even global in scope. They vary in their
focus from single categories of organisms—for example,
amphibians or ants—to all categories of life.

As the information comes together online, the big picture of
Earth's biodiversity will emerge as a mosaic at high resolution.
Despite their modest outward appearance, the all-species inven-

tories are in fact collectively "big" science, a moonshot effort that will eventually engage many times the number of professional and citizen scientists now active. The positive impact of this scientific knowledge on medicine, agriculture, and resource management will be beyond measure. It will collaterally establish a foundation for the universal conservation of species and locally adapted genetic races. What is to be learned will at last reveal the full magnitude of the Creation.

V
REACHING ACROSS

SCIENCE AND RELIGION

ARE THE TWO MOST POWERFUL FORCES

OF SOCIETY. TOGETHER THEY

CAN SAVE THE CREATION.

17

An Alliance for Life

PASTOR, I am grateful for your attention. As a scientist who has spent a lifetime studying the Creation, I have done my best here to brief you and others on subjects I hope will be more part of our common concern. My foundation of reference has been the culture of science and some of secularism based on science, as I understand them. From that foundation I have focused on the interaction of three problems that affect everyone: the decline of the living environment, the inadequacy of scientific education, and the moral confusions caused by the exponential growth of biology. In order to solve these problems, I've argued, it will be necessary to find common ground on which the powerful forces of religion and science can be joined. The best place to start is the stewardship of life.

Obviously, neither religion nor science has addressed this great issue effectively. I've attempted to identify those elements of biology and education most relevant to the proposed partnership. In the process I've not tried to water down in any way the fundamental difference between science and mainstream religion concerning the origin of life. God made the Creation, you say. This truth is plainly stated in Holy Scripture. Twenty-

165

five centuries of theology and much of Western civilization have been built upon it. But no, I say, respectfully. Life was self-assembled by random mutation and natural selection of the codifying molecules. As radical as such an explanation may seem, it is supported by an overwhelming body of interlocking evidence. It might yet prove wrong, but year by year that seems less probable. And it raises this theological question: Would God have been so deceptive as to salt the earth with so much misleading evidence?

Much as I would like to think otherwise, I see no hope for compromise in the idea of Intelligent Design. Simply put, this proposal agrees that evolution occurs but argues that it is guided by a supernatural intelligence. The evidence for Intelligent Design, however, consists solely of a default argument. Its logic is simply this: biologists have not yet explained how complex systems such as the human eye and spinning bacterial cilium could have evolved by themselves; therefore a higher intelligence must have guided the evolution. Unfortunately, no positive evidence exists for Intelligent Design. None has been proposed to test it. No theory has been suggested, or even imagined, to explain the transcription from supernatural force to organic reality. That is why statured scientists, those who have led in original research, unanimously agree that the theory of Intelligent Design does not qualify as science.

Some have suggested that scientists have formed a conspiracy to halt the search for Intelligent Design. There is no such conspiracy. There is only agreement among experts that the hypothesis has none of the defining qualities of science. To think otherwise is to misunderstand the culture of science. Discoveries and the testing of discoveries are the currency of science, its irreplaceable silver and gold. Challenges to prevailing theory on the basis of new evidence is the hallmark of science. If positive and repeatable evidence were adduced for a supernatural intelligent force that created and guided the evolution of life, it would deservedly rank as the greatest scientific discovery of all time. It would transform philosophy and change the course of history. Scientists dream of making a discovery of this magnitude!

Without such an event, however, it is a dangerous step for theologians to summon the default argument of Intelligent Design as scientific support for religious belief. Biologists are explaining the previously unexplainable—providing evolutionary steps for the autonomous origin of ever more complex systems—at an accelerating pace. What is to become of the hypothesis of Intelligent Design as the remaining unpenetrated systems decline toward the vanishing point? The hypothesis will be dismissed, and with it credibility of the idea of science-based theology. The odds powerfully favor such an outcome. In science, as in logic, a default argument can never replace positive evidence, but even a sliver of positive evidence can demolish a default argument.

You and I are both humanists in the broadest sense: human welfare is at the center of our thought. But the difference between humanism based on religion and humanism based on science radiates through philosophy and the very meaning we assign ourselves as a species. They affect the way we separately authenticate our ethics, our patriotism, our social structure, our personal dignity.

What are we to do? Forget the differences, I say. Meet on common ground. That might not be as difficult as it seems at first. When you think about it, our metaphysical differences have remarkably little effect on the conduct of our separate lives. My guess is that you and I are about equally ethical, patriotic, and altruistic. We are products of a civilization that rose from both religion and the science-based Enlightenment. We would gladly serve on the same jury, fight the same wars, sanctify human life with the same intensity. And surely we also share a love of the Creation.

In closing this letter, I hope you will not have taken offense when I spoke of ascending to Nature instead of ascending away from it. It would give me deep satisfaction to find that expression as I have explained it compatible with your own beliefs. For however the tensions eventually play out between our opposing worldviews, however science and religion wax and wane in the minds of men, there remains the earthborn, yet transcendental, obligation we are both morally bound to share.

Warmly and respectfully,
Edward O. Wilson

References and Notes

Page

15 **The concepts of Nature and wilderness**, especially as cultural con-
structions, are examined in detail by the perspectives of many scholars
in William Cronon, ed., *Uncommon Ground: Toward Reinventing Nature*
(New York: W. W. Norton, 1995); and with particular reference to
American cultural history by Roderick Nash in *Wilderness and the American Mind*, 4th ed. (New Haven: Yale University Press, 2001). The concept of wilderness from scientific evidence is reviewed by Edward O.
Wilson, *The Future of Life* (New York: Alfred A. Knopf, 2002). A recent
critique of the constructivist view, among those of many authors, is in
Eileen Crist, "Against the Social Construction of Nature and Wilderness," *Environmental Ethics* 26 (2004): 5–24.

18 **On the Boston Harbor Islands**, see Charles T. Roman, Bruce Jacob-
son, and Jack Wiggin, "Boston Harbor Islands National Park Area:
Natural Resources Overview," special issue 3, *Northeastern Naturalist* 12
(2005): 3–12.

26 **On the value of wild Nature**, the technical and popular literature is
immense. I have reviewed many of the key aspects in my trilogy *The
Diversity of Life* (Cambridge: Harvard University Press, 1992), *Consilience: The Unity of Knowledge* (New York: Alfred A. Knopf, 1998), and
The Future of Life (New York: Alfred A. Knopf, 2002).

39 **The English translation of Fray Bartolomé de Las Casas** is by Sandra
Ferdman in *The Oxford Book of Latin American Short Stories*, ed. Roberto
González Echevarría (New York: Oxford University Press, 1997).

62 **Nature appreciation in the mid-1800s**: George Catlin, *Letters and Notes on the Manners, Customs, and Condition of the North American Indians*, vol. 1 (London, 1841), pp. 260–64.

63 **Biophilia**: in the growing literature, see Edward O. Wilson, *Biophilia* (Cambridge: Harvard University Press, 1984); Stephen R. Kellert and Edward O. Wilson, eds., *The Biophilia Hypothesis* (Washington, D.C.: Island Press/Shearwater Books, 1993); and Stephen R. Kellert, *Kinship to Mastery: Biophilia in Human Evolution and Development* (Washington, D.C.: Island Press, 1997).

63 **The new academic disciplines** of environmental psychology and conservation psychology are described by Carol D. Saunders, "The Emerging Field of Conservation Psychology," *Human Ecology Review* 10 (2003): 137–49.

66 **The principle of preferred habitat** in humans was developed by George H. Orians and Judith H. Heerwagen, "Evolved Responses to Landscapes," in Jerome H. Barkow, Leda Cosmides, and John Tooby, eds., *The Adapted Mind: Evolutionary Psychology and the Generation of Culture* (New York: Oxford University Press, 1992).

69 **The importance of natural settings to mental health** is reviewed by Howard Frumkin, "Beyond Toxicity: Human Health and the Natural Environment," *American Journal of Preventive Medicine* 20 (2001): 234–40.

73 **A general account of the extinction process** is given in my *The Diversity of Life* and *The Future of Life*.

77 **The decline of terrestrial, freshwater, and marine ecosystems** is documented by Jonathan Loh and Mathias Wackernagel, eds., in *Living Planet Report 2004* (Gland, Switzerland: WWF-Worldwide Fund for Nature, 2004).

78 **The decline of coral reefs** over most of the world is documented by D. R. Bellwood, T. P. Hughes, C. Folke, and M. Nyström, "Confronting the Coral Reef Crisis," *Nature* 429: (2004) 827–33.

79 **The decline of amphibians** is detailed by Simon N. Stuart et al., "Status and Trends of Amphibian Declines and Extinctions Worldwide," *Science* 306 (2004): 1783–86. I am grateful to James Hanken for supplying recently completed data on the status of the Haitian frogs.

87 **The finding of the ivory-billed woodpecker** and the list of American bird species extinguished since 1980 are reported by David S. Wilcove

in "Rediscovery of the Ivory-billed Woodpecker," *Science* 308 (2005): 1422–23.

92 **The bottleneck imagery** was developed in detail in my *Consilience* and *The Future of Life*.

93 **Signatories to the Convention on Biological Diversity**, and their targets for abating extinction, are cited by Thomas Brooks and Elizabeth Kennedy, "Conservation Biology: Biodiversity Barometers," *Nature* 431 (2004): 1046–48.

93 **National constitutions with provisions to protect Nature** are reviewed by David W. Orr in "Law of the Land," *Orion*, January/February 2004, pp. 18–25.

93 **Species loss during the next half century** as a result of global warming alone is the estimate of Chris T. Thomas et al., "Extinction Risk from Climate Change," *Nature* 427 (2004): 145–48. Also, see the commentary by J. Alan Pounds and Robert Puschendorf, "Ecology: Clouded Futures," ibid., 107–9.

95 **The thirty-four hot spots** are analyzed by Russell A. Mittermeier et al. in *Hotspots Revisited: Earth's Biologically Richest and Most Endangered Terrestrial Ecosystems* (Mexico City: Cimex, 2005).

98 **Marine conservation**, science and practice, especially with reference to the open seas, is presented by multiple authors in Linda K. Glover and Sylvia A. Earle, eds., *Defying Ocean's End: An Agenda for Action* (Washington, D.C.: Island Press, 2004).

99 **The size of marine reserves needed globally** and the estimated costs of their protection are given by Andrew Balmford et al., "The Worldwide Costs of Marine Protected Areas," *Proceedings of the National Academy of Sciences, USA* 101 (2004): 9694–97, and discussed by Henry Nicholls in "Marine Conservation: Sink or Swim," *Nature* 432 (2004): 12–14.

111 **Structure of DNA:** James D. Watson and Francis H. C. Crick, "A Structure for Deoxyribose Nucleic Acid," *Nature* 171 (1953): 737.

119 **The description here of the Encyclopedia of Life** project is drawn with modification from my article "The Encyclopedia of Life," *Trends in Ecology & Evolution* 18 (2003): 77–80.

139 **The development of a naturalist**: Much of what I know is drawn from my own experience and that of my close friends, recounted in my memoir *Naturalist* (Washington, D.C.: Island Press, 1994). But others

have written with the same feeling and more detail, including, for example, Richard Louv in *Last Child in the Woods: Saving Our Children from Nature-Deficit Disorder* (Chapel Hill, N.C.: Algonquin Books of Chapel Hill, 2005).

141 **The definition of naturalist intelligence** is given by Howard Gardner in *Intelligence Reframed: Multiple Intelligences for the 21st Century* (New York: Basic Books, 1999), pp. 49–50.

144 **The tendency to create hideaways** is analyzed by David T. Sobel in *Children's Special Places: Exploring the Role of Forts, Dens, and Bush Houses in Middle Childhood* (Tucson: Zephyr Press, 1993).

150 **The new species** discovered in the Great Smoky Mountains National Park by the All Taxa Biodiversity Inventory (ATBI) in early 2004 are listed in the *ATBI Quarterly*, Summer 2004, p. 3.

150 **David Wagner on the Lepidoptera inventory** of the Great Smoky Mountains National Park: "Results of the Smokies 2004 Lepidoptera Blitz," *ATBI Quarterly*, Summer 2004, pp. 6–7.

152 **The prospects of accelerating taxonomic research** and creating an electronic encyclopedia of all life forms are described by Edward O. Wilson in "On the Future of Conservation Biology," *Conservation Biology* 14 (2000): 1–3; and "The Encyclopedia of Life," *Trends in Ecology & Evolution* 18 (2003): 77–80.

154 **The account of the first Biodiversity Day**, in Massachusetts, is from my *The Future of Life*. Peter Alden supplied the list of states that have had bioblitzes, as Biodiversity Days are usually called now, and Ines Possemeyer supplied the list of other countries that are holding them in 2005 (personal communications).

155 **The bioblitz in New York City's Central Park**: Richard C. Wiese and Jeff Stolzer, "Exploring New York's 'Backyard,'" *Explorers Journal*, Summer 2003, pp. 10–13.

159 **Butterfly discoveries in the Dominican Republic**: Brian D. Farrell, "From Agronomics to International Relations," *Revista* (Harvard Review of Latin America Studies), Fall 2004/Winter 2005, pp. 7–9.

165 **Faith-based support for stewardship of the environment**, including to increasing degree biodiversity conservation, is cropping up in many religions and denominations around the world. In the United States, initiatives are being sponsored, for example, by the National Council of Churches, the National Religious Partnership for the Environment,

Presbyterians for Restoring Creation, the U.S. Conference of Catholic Bishops, and the Pacific Conference of the Methodist Church. A review of additional movements, including those in other countries and major religions, is provided by Jim Motovalli in "Steward of the Earth," *Environmental Magazine* 13, no. 6 (2002): 1–16. Especially notable among religious leaders is Patriarch Bartholomew, the "Green Patriarch," leader of the 300 million Orthodox Christians.

About the Author

BORN IN BIRMINGHAM, ALABAMA, in 1929 and raised a Southern Baptist in Alabama, Edward O. Wilson was lastingly influenced by the lyrical and spiritual power of evangelical Christianity. He was equally imprinted to the beauty and mystery of the natural environments he explored as a boy. These formative influences combined to draw him into evolutionary biology while a student at the University of Alabama. Scientific humanism, as explained in the present work, thereafter provided his organizing worldview, but he was never severed from his roots.

Wilson's career was focused on scientific research and teaching during graduate training and forty-one years on the faculty of Harvard University, and subsequent retirement. His twenty books and more than four hundred mostly technical articles have won him over one hundred awards in science and letters, including two Pulitzer Prizes, for *On Human Nature* (1978) and, with Bert Hölldobler, *The Ants* (1990); the U.S. National Medal of Science; the Crafoord Prize, given by the Royal Swedish Academy of Sciences for fields it does not cover by the Nobel

Prize; Japan's International Prize for Biology; the Presidential Medal and Nonino Prize of Italy; and the Franklin Medal of the American Philosophical Society. For his contributions to conservation biology, he has received the Audubon Medal of the National Audubon Society and gold medal of the Worldwide Fund for Nature. Much of his personal and professional life is chronicled in the memoir *Naturalist,* which won the *Los Angeles Times* Book Award in Science in 1995.

Still active in field research, writing, and conservation work, Wilson lives with his wife, Irene, in Lexington, Massachusetts.